Alexander Weddemann

Magnetic beads for microfluidic lab-on-a-chip devices

Alexander Weddemann

Magnetic beads for microfluidic lab-on-a-chip devices

A finite element model of a total analysis system: from transport and separation to positioning and magnetoresistive detection

Südwestdeutscher Verlag für Hochschulschriften

Impressum/Imprint (nur für Deutschland/ only for Germany)
Bibliografische Information der Deutschen Nationalbibliothek: Die Deutsche Nationalbibliothek verzeichnet diese Publikation in der Deutschen Nationalbibliografie; detaillierte bibliografische Daten sind im Internet über http://dnb.d-nb.de abrufbar.

Alle in diesem Buch genannten Marken und Produktnamen unterliegen warenzeichen-, marken- oder patentrechtlichem Schutz bzw. sind Warenzeichen oder eingetragene Warenzeichen der jeweiligen Inhaber. Die Wiedergabe von Marken, Produktnamen, Gebrauchsnamen, Handelsnamen, Warenbezeichnungen u.s.w. in diesem Werk berechtigt auch ohne besondere Kennzeichnung nicht zu der Annahme, dass solche Namen im Sinne der Warenzeichen- und Markenschutzgesetzgebung als frei zu betrachten wären und daher von jedermann benutzt werden dürften.

Verlag: Südwestdeutscher Verlag für Hochschulschriften GmbH & Co. KG
Dudweiler Landstr. 99, 66123 Saarbrücken, Deutschland
Telefon +49 681 37 20 271-1, Telefax +49 681 37 20 271-0
Email: info@svh-verlag.de
Zugl.: Bielefeld, Universität Bielefeld, Dissertation, 2009

Herstellung in Deutschland:
Schaltungsdienst Lange o.H.G., Berlin
Books on Demand GmbH, Norderstedt
Reha GmbH, Saarbrücken
Amazon Distribution GmbH, Leipzig
ISBN: 978-3-8381-1867-3

Imprint (only for USA, GB)
Bibliographic information published by the Deutsche Nationalbibliothek: The Deutsche Nationalbibliothek lists this publication in the Deutsche Nationalbibliografie; detailed bibliographic data are available in the Internet at http://dnb.d-nb.de.

Any brand names and product names mentioned in this book are subject to trademark, brand or patent protection and are trademarks or registered trademarks of their respective holders. The use of brand names, product names, common names, trade names, product descriptions etc. even without a particular marking in this works is in no way to be construed to mean that such names may be regarded as unrestricted in respect of trademark and brand protection legislation and could thus be used by anyone.

Publisher: Südwestdeutscher Verlag für Hochschulschriften GmbH & Co. KG
Dudweiler Landstr. 99, 66123 Saarbrücken, Germany
Phone +49 681 37 20 271-1, Fax +49 681 37 20 271-0
Email: info@svh-verlag.de

Printed in the U.S.A.
Printed in the U.K. by (see last page)
ISBN: 978-3-8381-1867-3

Copyright © 2011 by the author and Südwestdeutscher Verlag für Hochschulschriften GmbH & Co. KG and licensors
All rights reserved. Saarbrücken 2011

Preface

Small magnetic particles have attracted a lot of interest during the last decades. We may think of a magnetic particle as small magnetic sphere which interacts with other magnetic objects nearby due to its magnetic stray field. Such particles can be manipulated by external (inhomogeneous) magnetic fields because of their permanent magnetic moment. In particular, if particles are dissolved in a liquid, their interaction with such fields provides a method to guide them in a fluid flow or to restrict their motion to a certain volume. Especially for medical applications, this seemed to be of high impact. The pioneering idea was to bind drug agents to such magnetic objects which could then serve as magnetic carriers allowing for the indirect control of the medical compounds. Such strategies could lead to strong advances for many treatments applied nowadays. In particular, they might help to reduce the side effects of non-localized treatments such as chemotherapy for cancer therapy where the drugs are only needed in a specific area of the body i.e. near to the malignant tumour cells. However, as no directed transport is possible via only employing the blood flow, therapeutic compounds necessarily travel through the entire body and only a small fraction reaches the actual target. Therefore, large quantities of drugs need to be administered which can lead to severe side effects. Guiding the combined particle/drug-system to or capturing it at the target region would reduce the dosage needed and allow for more efficient treatments (Dobson, 2006; Langer, 1990, 1998; Gupta and Hung, 1989, 1990; Gallo et al., 2006). In this regard, magnetic particles have been

Preface

thoroughly studied in respect to the possibility of surface modifications which enable their binding to such drug agents (Gao et al., 2005; Vadala et al. 2005). A strong localization is, in particular, required when applying magnetic particles in hyperthermia. Here, particles remain stationary in a high frequency (up to 100 kHz, but homogeneous) magnetic field close to the tumour. Their magnetic moment vector rotates together with the field direction and, due to microscopic damping phenomena, the particles serve as heat sources of the surrounding area. A temperature increase up to ~ 46°C has been reported to destroy many tumour cells (Nader, 2009; Salloum et al., 2008; Jordan et al., 2009). Other applications are the employment of magnetic particles as markers detectable by magnetoresistive sensors (Ferreira et al., 2003; Brezska et al., 2004, Schotter et al., 2003; Graham et al. 2004), due to their relaxation (Möller et al., 2005) or in an indirect manner: superparamagnetic colloids improve the contrast in Magnetic Resonance Imaging (Shultz et al., 2007; Magin et al., 1991; Coroiu et al., 2005; Winter et al., 2005).

Though during the last years a lot of effort from experimental as well as theoretical side was done and for many applications a estimates on applicability and limitations have been obtained, one more obstacle still needs to be tackled in the next years to open such strategies the way for the actual in vivo-application: The proof of their biocompatibility is still missing or, in other words, it is not accepted at present that magnetic micro- or nanoparticles are not toxic. Numerous of studies have been carried out in this field (Mahmoudi et al., 2009a, b; Macaroff et al., 2005), however, a conclusive result is yet to be obtained. Therefore, magnetic particles are still restricted to laboratory tasks.

In this regard, one specific "laboratory task" has developed rapidly in the last years: the *lab-on-a-chip* technology, which aims for the integration of all laboratory procedures on a small chip (Edelstein et al., 2000; Jiang et al., 2006; Pamme et al., 2006a). Due to the miniaturization, this device may be employed with only a small sample volume and can be integrated into portable devices allowing for applications in any area of the world. The latter advantage was the main driving force when, about three-and-a-half years ago, a consortium of about ten different participants came together for the first time in the small town of Bernried located directly on the wonderful shores of the Starnberger Lake. The reason for this gathering was the "kickoff" meeting of the BMBF project "MrBead" sponsored by the DFG (whose financial support ma be gratefully acknowledged at this point again). The focus of this project was the development of a portable, hand-held device for fast, reliable testing of liquids in respect to certain components. Such a device allows for numerous applications, ranging from infection or drug abuse testing in human diagnostics to food testing and also veterinary medicine. A constraint in this project was that the actual recognition needed to be achieved by magnetic particles. It may be interesting to remark that at same time the company Philips® (Megens and Prins, 2005) worked on a similar approach.

What is the idea of employing magnetic particles for the recognition of biological molecules? Due to their magnetic stray field, magnetic particles influence other magnetic objects nearby. In particular, this allows their detection via magnetoresistive sensors. For the realization of the detection, particles and sensor surfaces are specifically coated by different biological surfactants. The surface modifications must be chosen so that the biomolecules to be detected serve as linkers between both components. Magnetic particles carrying the proper type of linker (e.g. antibodies in human diagnostics) can thus bind to the surface. Due to the biological bond, they withstand a washing process which serves to purify the sensor surface from sedimented markers after the binding has taken place and thus may afterwards be detected. This indirectly proves the existence of the linker. From the diagnostic point of view, several questions arise at this point:

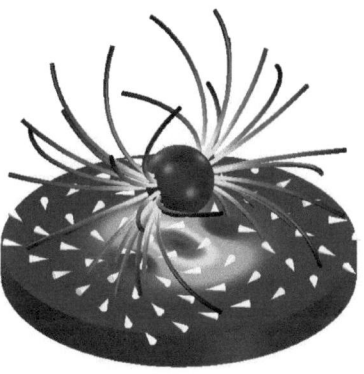

Schematic representation of a circular disc under the influence of a magnetic particle, lines show the particle field, arrows the components within the disc plane.

a) What exactly is measured? Can we distinguish whether 1 particle is placed on the sensor or more? Or do we need a certain threshold number of particles unless there is no signal? Can we conclude from the signal strength the number specific bounds and maybe obtain an estimation of the linker concentration in the original sample? In principle, the fundamental question is what can be

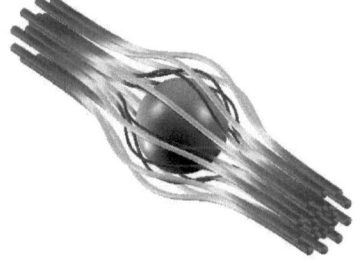

Particle dissolved in a fluid flow, the (stream)lines indicate the behaviour of the surrounding liquid.

iii

Preface

measured and how does the measured signal correspond to an infection or a lack of infection.

b) How do particles get to the sensor? Is it sufficient to wait for a while and let a combined influence of diffusion and gravity work or does this approach require too much time? Can we pull particles close the sensor by e.g. employing magnets, either permanent or by switchable current densities? These questions centre on the design of the whole internal structure of the device, the microfluidic channel geometry, also focusing on the time scale that is necessary to guarantee that all procedures can take place.

c) Can we keep it simple and fast? The device should be mainly designed for fast, reliable testing employed by non-specialists. Therefore, no complex handling of the sample was allowed. Ideally, the sample is just deposited into the device and after a short time of about 5 to 10 minutes a result should be given.

In the summer of 2006, these were the main questions raised, maybe not yet in such a specific form, but in principle very similar. One fact, however, was already decided: the device would consist of two components, the *handheld* and the *disposable*. The handheld consists of all those parts which never come into contact with the injected samples: such as data evaluation components, energy source or control keys to start the measuring procedure. A liquid sample would only flow through the disposable which thus needs to be replaced after every measurement. It contains the actual detection site with microfluidic guidance structure, sensor arrays and contacts to the handheld. In principle, its tasks can be summarized as follows:

a) bring particles into solution and enable the specific binding of molecules that shall be detected to the magnetic markers
b) transport particles to the detection site, maybe separate different products if multiple tests are evaluated at the same time
c) detect particles and conclude whether certain components are in the sample or not

Therefore, different components were to be discussed. On the one hand, we needed to design a microfluidic geometry for the transport of particles through different functional sites of the device. On the other hand, the electromagnetic properties of particles, manipulating external fields and the magnetoresistive sensors needed to be understood. Especially in the case of the detection properties, many fundamental questions arose. In the framework of this thesis, these questions were analyzed by means of simulations. The equations for the description of different components are based on continuum theories and are therefore given by partial differential equations. These equations were solved by finite element methods which have become one of the most popular numerical methods when dealing with such systems. Finite elements work by means of the reformulation of the original equation into the so-called *weak form* which may be regarded as a generalization in the sense that a solution of the original problem always solves the weak form but not necessarily vice versa. These weak or variational formulations can be discretized in a finite dimensional subspace of the original solution space which leads to a set of linear equations $Ax = b$ solvable by standard methods of numerical linear algebra. A brief summary of how a partial differential equations can be discretized by these ideas is given in chapter 1.

The second chapter deals with the setup of the microfluidic channel design. We will give a short introduction to the theory of hydrodynamics focusing on the case of microfluidic which is governed by highly laminar flows. Applying hydrodynamic, electromagnetic and gravitational forces on the magnetic carriers, we will introduce a lab-on-a-chip system which can handle injection, reaction, separation, positioning, and detection procedures. We will prove that the proposed geometry maintains certain transport properties for specific time scales. Despite the many different tasks, each component employs only a small number of physical phenomena. In particular, the amount of components on the microscale such as small current leading wire geometries is minimized. Therefore, each part has high potential to lower the complexity of existing lab-on-a-chip devices.

The tools for the discussion of magnetic phenomena are briefly reviewed in chapter 3. In particular, the fundamentals of micromagnetic calculations will be explained. These will afterwards be applied for the understanding of superparamagnetic particles and their interactions in microfluidic channels (chapter 4) and to the description of the magnetoresistive detection (chapter 5).

For all the simulations carried out in the framework of the thesis as well as for most of the visualizations shown in the figures, COMSOL Multiphysics^{TM}, 2005, was

Preface

used. For a short introduction on how weak equations are implemented refer to Appendix A.1.

Since the goal of the project is to 100% applied, all predictions important for either handheld or disposable had to be tested in experiments. Experiments on the microfluidic devices were carried out by F. Wittbracht, 2007, 2009 and B. Eickenberg. The experimental investigation of particle detection by magnetoresistive tunnelling sensors was the topic of the PhD thesis of C. Albon, 2009.

Acknowledgements

I guess, everybody participating in the project "PhD thesis" will make similar experiences. Going through all different emotional stages from disheartening desperation, constantly in the believe that everybody knows and understands so much more than oneself and that nothing will ever work as it is supposed to, to gushing enthusiasm, once things finally start giving results, one finally comes to the point realizing that there are (and always were) a certain number of "constant" people involved who contributed in various ways to the success of this journey. Now, looking back to these very exciting three years with some distance, I would like thank these people for their continuous scientific and moral support.

First of all, I would like to thank my supervisor Prof. Dr. Andreas Hütten for his advice and mentorship during the past years and for providing me with the possibility to work in this very interesting topic in the frame of the university-industry joint project MrBead. Especially, I would like to thank him for believing in the success of this work when no one else seemed to. I would also like to thank Prof. Dr. Günter Reiss for his support over the years as well as Dr. Simone Herth and Dr. Michael Schilling for introducing me to the subject of microfluidic devices.

Another very big "Thanks" goes to all those who carried out the experiments on the systems studied in this work and, therefore, significantly contributed to the success of this project. In particular, I would like to thank Frank Wittbracht who has realized most of the microfluidic devices under the assistance of the 'lost son' Bernhard Eickenberg who hopefully returns to our group at the beginning of next year. For providing me with many experimental measurements on the magnetoresistive particle detection and, thus, guiding many of the theoretical strategies in long and fruitful discussions, I would like to thank Camelia Albon and Peter Hedwig. I would like to thank Alexander Auge who was of great help in the conception and establishment of different aspects of the micromagnetic simulations and also Daniel Kappe for his assistance during the simulations of the interacting magnetic particles.

I can say for sure, that when I started this work several years ago, my mathematical background would not have been sufficient to understand the convergence difficulties or the numerical instabilities which arise in the models discussed. Therefore, I would like to thank V. Thümmler and W.-J. Beyn from the Bielefeld department of mathematics who provided me with a lot of fundamental knowledge on finite element schemes in ALE-based methods. The strategies and methods I learnt during my diploma thesis in mathematics have strongly contributed to the success of the numerical models applied in this work.

I would also like to thank the MrBead-partners, in particular, MicroCoat who provided biofunctionalized particles and surfaces for the analysis of the capturing properties of the microfluidic structure, and Reiner and I-Sys who were of great help for the realization and the design of several microfluidic components.

Special thanks go to my office mates Irina Dück and Markus Schäfers, who have endured me during the last years, Dieter Akemeier as well as N. Mill, A. Dreyer and Michael Peter because the time would have been a lot less fun without you guys. I would also like to thank all the D2-members for the nice working atmosphere during the last couple years. Further, I would like to thank Alethea Wait for intensive proofreading of the manuscript and Prof. Dr. J. Schnack for writing the second report on this thesis.

Finally, I thank my family who provided me with not only financial but also moral support during the last seven years of studying.

To my grandfather

Table of contents

1. Finite element modelling — 1
  1.1 Basic notations 5
  1.2 A short introduction to finite element methods 8
    1.2.1 Weak formulation 8
    1.2.2 Galerkin discretization 10
    1.2.3 Domain triangulation and finite elements 11
    1.2.4 Assembly and stiffness matrix 14
    1.2.5 Parabolic equations and time integration 15
  1.3 Living on a bubble – Moving domains 18
    1.3.1 Level-set-method 19
    1.3.2 ALE-method... 20

2. Particles in microfluidic devices — 25
  2.1 Fundamental of hydrodynamics 28
    2.1.1 Continuum hypothesis and effective parameters 29
    2.1.2 Lagrange- and Eulerian frame 32
    2.1.3 Navier-Stokes equation and Reynolds number 33
    2.1.4 The special case of microfluidics 35
    2.1.5 Spherical objects dissolved in liquids 37
    2.1.6 Boundary conditions 40
    2.1.7 Weak formulation 41
    2.1.8 Numerical stabilization and Petrov-Galerkin discretization 42
  2.2 Particle separation by a hydrodynamic switch 45
  2.3 Transport properties for the low Péclet-regime 51
  2.4 Microfluidic gravitational positioning system 58
  2.5 Conclusions 65
    2.5.1 Outlook 66

3. Magnetism — 69
  3.1 From atomic to mesoscopic magnetism 70
  3.2 Coupling between mesoscopic moment and atomic structure 72
  3.3 Magnetostatics in matter 74
  3.4 Static micromagnetism 75

Table of contents

    3.5 Dynamic micomagnetism 78

4. Magnetically interacting particles   81
    4.1 Superparamagnetism 82
    4.2 Homogenously magnetized particles.. 84
    4.3 Magnetization dynamics 85
    4.4 Dipolar driven demagnetization processes 88
    4.5 Magnetic particles in adiabatically changed magnetic fields 94
    4.6 Conclusion and Outlook 101
        4.6.1 Outlook: Magnetic particles in high frequency magnetic fields 102

5. Detection of magnetic particles   109
    5.1 Weak formulation of the thin film approach 111
        5.1.1 Tunnelling magnetoresistance sensors 114
        5.1.2 COMSOL implementation: PADIMA 116
    5.2 Manipulation of magnetic vortex states 118
    5.3 Space resolutive magnetic detection: "magnetic lenses"..................... 120
        5.3.1 Comparison to experimental data 120
        5.3.2 Estimation of spatial resolution limits 124
        5.3.3 Sensors for continuous flow particle measurements..................... 129
    5.4 Number resolutive magnetic detection 133
    5.5 Conclusion and Outlook 136

6. A MrBead-summary   139

Appendix   141
    A.1 COMSOL Multiphysics 141
    A.2 Magnetic point-particle under an external force 144
    A.3 Variational derivation of the free energy functional
            for micromagnetism 145
    A.4 Short introduction to PADIMA 150

List of references   160

Chapter 1

Finite element modelling

Many physical phenomena introduce observables depending on time and space: A cup of hot coffee loses its temperature while the surrounding area gets warmer, a stone thrown into a river induces a surface wave travelling spherically away from the point where the stone hit the liquid and a leaf is blown through the air along chaotic paths by the wind. All these phenomena can be described by partial differential equations (PDEs) which are equations for a set of dependent variables (in the examples temperature, surface displacement / liquid velocity and wind velocity) consisting of not only the function values but also of the partial derivatives of a certain order. In this work, the following examples will prove the most important: (a) the *advection-diffusion equation* describing the dynamics of a particle concentration under the influence of thermal motion as well as an additional convection field. Such fields are often obtained from (b) the *Navier-Stoke/Stokes equation* which together with the *equation of continuity* is the governing equation for the evolution of a fluid flow. On the electromagnetic side, we will need to solve (c) the *Landau-Lifshitz equation* combined with (d) the *Maxwell equations* for matter whose solutions describe the magnetization dynamics in magnetic systems.

Except for very simple systems, the equations mentioned above cannot be solved by analytic methods but rather their solutions need to be approximated by numerical means. To get a better impression of what we are talking about, we will consider a seemingly easy example: Let be $\Omega \subset \mathbb{R}^n$ and denote by Δ the Laplace operator. For a

1. Finite elements

continuous mapping $f : \Omega \to \mathbb{R}$ ($f \in C(\Omega)$) the *inhomogeneous Laplace / Poisson problem* is to find a sufficiently smooth function $u : \Omega \to \mathbb{R}$ ($u \in C^2(\Omega)$) so that

$$\Delta u = f. \tag{1.1}$$

This equation can be found in many physical fields, e.g. in electrostatics with $u = \phi$ the electric potential and $f = \rho$ the electric charge density. Similar to the way in which a time interval I must be specified when defining ordinary differential equations (ODEs), when working with PDEs an area $\Omega \subset \mathbb{R}^n$ needs to be chosen. If Ω is bounded, additional boundary conditions are necessary. From a mathematical point of view, the properties of solutions of (1.1) are well understood, for the unbounded case as well as for bounded geometries (Larsson and Thomée, 2005):

Considering a boundary problem with data f either $f \leq 0$ or $f \geq 0$ on Ω, solutions follow a *maximum principle*, i.e. they reach their respective maximum or minimum at the boundary. Such a property may be readily employed to prove certain stability estimations ensuring that under a small variation of the data f only a small variation of the solution u is expected. In the unbounded case, explicit solutions can be found by first constructing a *fundamental solution* which is a function $U : \Omega \to \mathbb{R}$ satisfying $\Delta U = \delta$ with δ the Kronecker δ-distribution. Solutions u for arbitrary data f can be obtained from the convolution $u = U * f$ as long as f is at least continuously differentiable and reaches all of its non-zero values in a bounded subset of Ω. For bounded geometries, a similar relation only holds for a specific type of fundamental solution, the so-called *Green's function* $G : \Omega \times \Omega \to \mathbb{R}$ fulfilling certain continuity conditions along the boundary. Solution u can be obtained according to

$$u(x) = \int_\Omega G(x, y) f(y) \, dy \tag{1.2}$$

Therefore, finding analytic solutions entails finding a fundamental solution or Green's function. Though such a solution would be ideal, this analytical approach is not typically used in mathematical analyses, as there is no known method to directly derive the Green's function. On the other hand, the existence of analytic solutions is considered unproven for many systems. The most famous example is probably the Navier-Stokes problem which has been nominated by the Clay Mathematics Institute, 2009, as one of the seven Millennium Prize Problems. Therefore, we will not stress this point further. Instead, we will pursue a different strategy.

The task is instead to find an approximate solution of equation (1.1) for given initial and boundary values by numerical means. The general method of dealing with this problem is actually very old and was first applied by Alexander Hrennikoff in 1941 (Hrennikoff, 1941): The domain Ω is divided into smaller subdomains. On each subdomain the approximate solution is constructed from an appropriate function space (that of course should be chosen to be as simple as possible) of a finite dimension. A finite dimension ensures that the functions are completely defined by a finite number of degrees of freedom, e.g. such as affine linear functions $g : \mathbb{R} \to \mathbb{R}$ are determined by their values at two points. Due to this discretization, it is possible to reduce the original problem to a system of linear equations $Ax = b$ which can be solved by standard methods of numerical mathematics (e.g. Newton iteration).

However, for a long time this method yielded severe complications. To understand why, in example (1.1) the function spaces have been given in which each component of the equation needs to be chosen. This was not done to confuse the reader or to make the example look more complex, but instead to underline the basic properties that need to be fulfilled so that u can be a solution of the equation. Figure 1.1 shows a schematic representation of how a numerical approximation to the solution might look: the exact solution (black line) was approximated by a linear spline (red line). However, such an approximation can no longer be a solution of the original equation because it is not even differentiable. In the example considered in section 1.2, we will

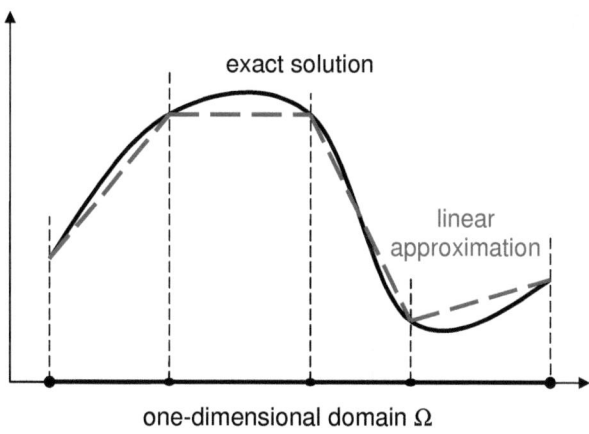

Figure 1.1: Exact solution (black) approximated by a piecewise linear, continuous function (red). Since the approximation is not differentiable itself, it cannot be a solution of the original system of partial differential equations.

1. Finite elements

show that it is possible by some reformulation of the original equation (1.1) to generalize the inhomogeneous Laplace equation to a formulation requiring u to be only once instead of twice partially continuously differentiable. Therefore, original strategies were pursued approximating the actual solution by e.g. piecewise trigonometric functions to maintain regularity. Yet, such approaches led to severe complications which restricted the application of such numerical schemes to only a small set of simple problems. The breakthrough was achieved when it was realized that the original problem could be recast into generalized function spaces only requiring the approximating functions to be continuous. The ideas and methods introduced and developed by Boris Galerkin, John William Rayleigh and Walther Ritz are still the fundamental basis of modern finite element methods, which have become one of the most powerful numerical methods to solve partial differential equations.

In this chapter, the basic ideas of finite element methods are demonstrated analyzing the inhomogeneous Laplace equation (1.1), as well as the corresponding parabolic system on static and moving domains. After introducing the most common function spaces necessary to understand this chapter in paragraph 1.1, the main steps for retrieving the discretized system from the original set of partial differential equations is explained in 1.2: We will recast the original equation into its corresponding *variational* or *weak form* (section 1.2.1). The weak formulation may be understood as a generalized version in the sense that it is well defined not only for sufficiently smooth functions, but also for mappings of less regularity. In a next step, the equation needs to be discretized by a *Galerkin approach* approximating the original solution spaces by finite-dimensional subspaces (section 1.2.2). Introducing the *Lagrangian finite elements*, we provide an explicit method with which to construct the linear space approximations (section 1.2.3). Since each individual basis function has a very small support (i.e. each basis function is different from zero only among a small subset of Ω), the system matrix A of the corresponding linear system $Ax = b$ has only a small amount of entries that are unequal to zero. This *sparsity* of the system matrix is one of the key advantages of finite element methods (section 1.2.4). The introductory section will close with a short review on parabolic equations (section 1.2.5).

The parabolic equation may be used to describe time-dependent problems. However, a certain type of time dependence cannot be treated in such a framework: a system where the domain Ω itself evolves in respect to time. Such phenomena are often encountered in the field of fluid structure interactions, where large deformations can be found which can no longer be incorporated by linear approaches. Two possible strategies will be introduced in 1.3: (a) the *Level-set-method*, which models interfaces between different geometries as an implicit function g as the root function of a higher dimensional mapping $\Phi(g) = 0$. (b) the *ALE-(Arbitrary Lagrangian Eulerian) method* which employs multiple frames to map the moving domain onto a non-moving configuration.

1.1 Basic notations

Finite element methods cannot directly deal with unbounded geometries as it is necessary to decompose the physical domain into a finite number of subdomains (as will be explained in section 1.2). Therefore, if not explicitly stated, $\Omega \subset \mathbb{R}^n$ with $n = 2$ or $n = 3$ denotes an open, bounded set with boundary $\partial \Omega$ and closure $\bar{\Omega}$. The set of continuous functions $u : \Omega \to \mathbb{R}$ on Ω will be denoted by $\mathcal{C}(\Omega)$. For the subspace of all k-times continuously differentiable functions, we write $\mathcal{C}^k(\Omega)$ and denote by $\mathcal{C}_0^k(\Omega)$ the subset of all functions with compact support on Ω. Furthermore, we make use of the standard notations for differential operators, defining the Nabla- and the Laplace-operator by $\nabla = (\partial_1, ..., \partial_n)^T$ and $\Delta = \partial_1^2 + ... + \partial_n^2$, respectively. As already seen in Figure 1.1, the approximating functions are usually only continuous but not differentiable. Therefore, it is necessary to consider function spaces of less regularity, e.g. the *p-integrable* functions $L^p(\Omega)$ and its subspaces, the *Sobolev spaces* $W^{k,p}(\Omega)$ and $H^k(\Omega)$. Detailed information on all definitions introduced in this chapter can be found in standard textbooks for functional analysis, e.g. Triebel, 1980; Amann and Escher, 2001; Hanke-Bourgeois, 2006. In this work, we will indicate by $L^p(\Omega), 1 \leq p < \infty$ the set of measurable functions $u : \Omega \to \mathbb{R}$ on Ω, so that

$$\int_\Omega |u|^p\, dx < \infty. \tag{1.3}$$

Equipping $L^p(\Omega)$ with the norm and additionally $L^2(\Omega)$ with the scalar product

$$\|u\|_{L^p(\Omega)} = \left(\int_\Omega |u|^p\, dx \right)^{1/p} \qquad u \in L^p(\Omega) \tag{1.4a}$$

1. Finite elements

$$\langle u,v\rangle_{L^2(\Omega)} = \sqrt{\int_\Omega uv\,dx} \qquad u,v \in L^2(\Omega) \tag{1.4b}$$

the set $L^p(\Omega)$ and $L^2(\Omega)$ form a Banach and a Hilbert space, respectively.

When discussing finite element methods, $L^2(\Omega)$, the space of *square-integrable functions*, is the most common function space. However, it is usually not sufficient to consider $u \in L^2(\Omega)$ but it is also necessary that partial derivatives up to an order k are $L^2(\Omega)$-functions. Unfortunately, definition (1.3) does *not* ensure any kind of regularity. In particular, functions in $L^p(\Omega)$ are usually not even continuous. As an example, one might take the step functions shown in Figure 1.2 which obviously have points of discontinuity but the integrals of $|u|^p$ on a bounded domain always remain finite. Instead of the common (strong) definition of a derivative, it is necessary to use a generalized form, the *weak derivative* or the *derivative in sense of distributions*.

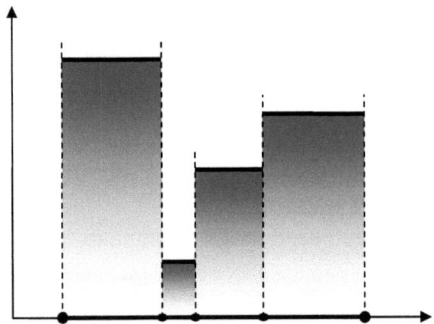

Figure 1.2: Step function. The function contains points of discontinuity and is therefore $\notin C(\Omega)$ but since $|\cdot|^p$ remains finite, it is $\in L^p(\Omega)$ for arbitrary p.

As a motivation, we choose $u \in C^1(\Omega)$ and $\psi \in C_0^\infty(\Omega)$ and follow by partial integration:

$$\int_\Omega \frac{\partial u}{\partial x}\psi\,dx = \int_{\partial\Omega} u\psi\,dx - \int_\Omega u\frac{\partial \psi}{\partial x}\,dx$$
$$= -\int_\Omega u\frac{\partial \psi}{\partial x}\,dx$$

For $L^p(\Omega)$-functions this identity is chosen as a definition:

Definition (weak derivative): Let $\Omega \subset \mathbb{R}^n$ be open, $\psi \in C_0^\infty(\Omega)$ and $f \in L^p(\Omega)$. The *weak derivative* is given by $g \in L^p(\Omega)$ such that

$$\int_\Omega g(x)\cdot\psi(x)\,dx = -\int_\Omega f(x)\frac{\partial \psi(x)}{\partial x}\,dx \qquad \forall \psi \in C_0^\infty(\Omega)$$

It stands that the weak derivative is always unique in the sense of $L^p(\Omega)$-functions, i.e. if g_1 and g_2 are two weak derivatives, they are identical except along a set of measure "0". Also, if f is differentiable in the strong sense, weak and strong derivative coincide. To give an example, we calculate the weak derivative of the heavy-side function Θ given by

$$\Theta(x) = \begin{cases} 0 & \text{für } x < 0 \\ 1 & \text{für } x \geq 0 \end{cases}$$

It is
$$\int_{\mathbb{R}} \frac{d\Theta(x)}{dx}\psi(x)\,dx \stackrel{\text{defintion}}{=} -\int_{\mathbb{R}} \Theta(x)\frac{d\psi(x)}{dx}\,dx = -\int_{0}^{\infty}\frac{d\psi(x)}{dx}\,dx$$
$$= -\lim_{x\to\infty}\psi(x) + \psi(0) = \psi(0) = \int_{\mathbb{R}} \delta(x)\cdot\psi(x)\,dx$$

Therefore, the weak derivative of the (discontinuous) heavy-side function Θ is given by the Kronecker δ-distribution.

Having introduced a definition for a differentiation that may be applied to $L^p(\Omega)$-functions, we are now able to introduce the *Sobolev spaces*. Denoting n-dimensional multiindices by $\alpha = (\alpha_1,\ldots,\alpha_n) \in \mathbb{N}_0^n$ and setting $|\alpha| := \alpha_1 + \ldots + \alpha_n$, we may write differential operators ∂^α via

$$\partial^\alpha := \frac{\partial^{|\alpha|}}{\partial x_1^{\alpha_1}\ldots\partial x_n^{\alpha_n}}$$

Definition (Sobolev space): For an integer $k > 0$ and $1 \leq p < \infty$, the *Sobolev space* $W^{k,p}(\Omega)$ is given by the subspace

$$W^{k,p}(\Omega) = \{u \in L^p(\Omega) \mid \partial^\alpha u \in L^p(\Omega)\ \forall \alpha \text{ with } |\alpha| \leq k\}$$

of $L^p(\Omega)$. In the special case of $p = 2$, we write $W^{k,2}(\Omega) = H^k(\Omega)$.

For the discussion of finite elements, it is often sufficient to restrict the analysis to $H^1(\Omega)$, to square-integrable functions with square-integrable first-order derivatives. $H^1(\Omega)$ becomes a Hilbert space with the scalar product

$$\langle u,v \rangle_{H^1(\Omega)} := \int_\Omega \langle \nabla u, \nabla v\rangle\,dx + \int_\Omega uv\,dx \qquad u,v \in H^1(\Omega) \qquad (1.5a)$$

Apart from the resulting norm $\|\cdot\|_{H^1(\Omega)} = \langle \cdot,\cdot\rangle_{H^1(\Omega)}^{1/2}$, a half-norm $|u|_{H^1(\Omega)}$ can be introduced by

1. Finite elements

$$|u|_{H^1(\Omega)} = \left(\int_\Omega |\nabla u|^2 \, dx \right)^{1/2} \qquad u \in H^1(\Omega) \qquad (1.5b)$$

Finally, it is necessary to have a generalization of the function space $C_0^\infty(\Omega)$ which is given by the closure of $C_0^\infty(\Omega)$ in $H^1(\Omega)$ in respect to the corresponding norm. We write

$$H_0^1(\Omega) = \overline{C_0^\infty(\Omega)}^{\|\cdot\|_{H^1(\Omega)}}$$

1.2 A short introduction to finite element methods

We now return to the example (1.1) considering a bounded domain Ω with smooth boundary $\partial\Omega$. To completely specify the problem, additional boundary conditions are necessary. For right now, we will assume homogeneous van Neumann conditions, i.e. along the boundary the derivative in direction of the surface normal vector $\hat{n}$ vanishes. The complete problem therefore is to find $u \in C^2(\Omega)$ such that

$$\Delta u(x) = f(x) \qquad \forall x \in \Omega \qquad (1.6a)$$
$$\langle \hat{n}, \nabla u \rangle = 0 \qquad \forall x \in \partial\Omega \qquad (1.6b)$$

with $f \in C(\Omega)$ and $\langle \cdot, \cdot \rangle$ the Euclidean inner product. For u to be a solution of this equation, u must be at least twice continuously differentiable; a solution satisfying this requirement is also called *strong solution* of (1.6). In particular, the piecewise linear approximation shown in Figure 1.1 does therefore not solve (1.6). The first step for applying finite element schemes to a partial differential equation is to lower the regularity constraints on a solution. Consequently, the new solution is not necessarily a solution of the original equation anymore but instead a generalized solution.

1.2.1 Weak formulation

When introducing the weak derivative in section 1.1, we "reassigned" the differential operation to a smooth function by partial integration. Here we proceed similarly: let be $\psi \in C^\infty(\Omega)$ an arbitrary smooth function or *testfunction*. If we multiply (1.6a) by ψ and integrate over Ω, we obtain

$$\int_\Omega \psi \Delta u \, dx = \int_\Omega \psi f \, dx.$$

Integrating the first summand by parts results in

$$\int_{\partial\Omega} \psi \langle \hat{n}, \nabla u \rangle \, dx - \int_\Omega \langle \nabla \psi, \nabla u \rangle \, dx = \int_\Omega \psi f \, dx.$$

If we finally exploit van Neumann conditions, we obtain the *variational formulation* of (1.6)

$$-\int_\Omega \langle \nabla \psi, \nabla u \rangle \, dx = \int_\Omega \psi f \, dx \qquad (1.7)$$

or $\quad a(\psi, u) = \ell(\psi) \quad$ with $\quad a(\psi, u) = -\langle \nabla \psi, \nabla u \rangle_{L^2(\Omega)}$
$$\text{and} \quad \ell(\psi) = \langle \psi, f \rangle_{L^2(\Omega)}$$

with $a : C^\infty(\Omega) \times C^2(\Omega) \to \mathbb{R}$ and $\ell : C^\infty(\Omega) \to \mathbb{R}$ a bilinear and linear mapping, respectively. Since we started with an arbitrary function $\psi \in C^\infty(\Omega)$, a solution u of (1.7) needs to satisfy the variational equation for *all* $\psi \in C^\infty(\Omega)$. However, it needs to pointed out that (1.7) only contains derivatives of first order of u. Therefore, a solution of (1.7) needs to meet a lower degree of regularity properties than a solution of (1.6). In this sense, (1.7) is *weaker* than (1.6). In particular, a solution of (1.6) is always a solution of (1.7) but the converse is not necessarily true. Usually certain conditions on the functions f need to be imposed, for examples see e.g. Hanke-Bourgeois, 2006. Due to van Neumann conditions, (1.7) does not contain any terms depending on the domain boundary; in this sense van Neumann boundary conditions are also referred to as *natural conditions* for finite element approaches. If we instead consider e.g. Dirichlet conditions, boundary integrals will remain in the variational form. To obtain a similar formulation to (1.7), the incorporation of boundary conditions is done by a proper choice of the testfunction space. Assuming $\psi \in C_0^\infty(\Omega)$ instead of $C^\infty(\Omega)$, boundary contributions always vanish; equation (1.7) now needs to hold $\forall \psi \in C_0^\infty(\Omega)$.

1. Finite elements

The question is now whether it is possible to find yet more general function spaces also containing functions which are easy to handle from a numerical point of view as e.g. piecewise polynomials. Obviously, it is not necessary that we choose $\psi \in C^{\infty}(\Omega)$ but all operations in (1.7) remain well-defined for $\psi \in C^{1}(\Omega)$. Indeed, understanding all derivatives in sense of weak derivatives, it is possible to show (Hanke-Bourgeois, 2006) that it is sufficient to have $u, \psi \in H^{1}(\Omega)$. Therefore, the *weak formulation* of (1.6) is given by

$$\text{find } u \in H^{1}(\Omega) \text{ such that}$$

$$a(\psi, u) = \ell(\psi) \qquad \forall \psi \in H^{1}(\Omega) \tag{1.8}$$

Solutions of (1.8) are also called *weak solutions*.

1.2.2 Galerkin discretization

By reformulating the original problem into the weak form, it was possible to decrease the degree of required regularity which a solution needs to meet. However, the function spaces $H_{0}^{1}(\Omega) \subset H^{1}(\Omega) \subset L^{2}(\Omega)$ are infinite dimensional linear spaces; a linear system of type $Ax = b$ still introduces an infinite number of degrees of freedom. Therefore, instead of trying to find an exact solution $u \in H^{1}(\Omega)$ of (1.8), we introduce a finite dimensional subspace $V_{h} \subset H^{1}(\Omega)$ from which approximate solutions u_{h} as well as approximate testfunctions ψ_{h} are constructed. The approximate problem may be written in a similar way to (1.8):

$$\text{find } u_{h} \in V_{h} \text{ such that}$$

$$a(\psi_{h}, u_{h}) = \ell(\psi_{h}) \qquad \forall \psi_{h} \in V_{h} \tag{1.9}$$

Since we consider a finite-dimensional V_{h}, we can choose a finite basis set $\{\Lambda_{1},...,\Lambda_{N}\}$ of V_{h}. Every approximate solution u_{h} of (1.9) may thus be written in the series expansion

$$u_h = \sum_{j=1}^{N} u_j \Lambda_j \qquad u_j \in \mathbb{R} \qquad (1.10)$$

Due to the linearity of the mappings $a(u_h, \cdot)$ and ℓ an approximate solution u_h satisfies (1.9) $\forall \psi_h \in V_h$ if it satisfies (1.9) on the finite basis set $\{\Lambda_1, ..., \Lambda_N\}$. Assuming the special form of a and ℓ deduced for the inhomogeneous Laplace problem, inserting (1.10) into (1.9) leads to the equation system

$$\ell(\Lambda_i) = a(\Lambda_i, u_h) = \int_\Omega \langle \nabla \Lambda_i, \nabla u_h \rangle \, dx = \sum_{j=1}^{N} u_j a(\Lambda_i, \Lambda_j)$$

which follows due to the linearity of $a(\cdot, \Lambda_i)$. Therefore, we may rewrite (1.10) in a system of linear equations. In matrix form, this is given by

$$AU = b \qquad (1.11)$$

with $U = (u_1, ..., u_N)^T$, $A = [a(\Lambda_i, \Lambda_j)]_{i,j} \in \mathbb{R}^{N \times N}$, $b = [\ell(\Lambda_j)]_j \in \mathbb{R}^N$.

This procedure is known as *Galerkin approach*. The matrix A is referred to as the *stiffness matrix*: it is a symmetric positive definite matrix which implies that equation (1.11) has a unique solution. Thus, the Galerkin approach always leads to a well-defined solution. For additional properties of the stiffness matrix see section 1.2.4.

1.2.3 Domain triangulation and finite elements

The preceding section explains how to reduce the original problem from an infinite- to a finite-dimensional problem. Formulation (1.11) may be readily used and solved e.g. by Newton iteration. The remaining unanswered question is how to construct the approximate solution space V_h or the set of basis functions $\{\Lambda_1, ..., \Lambda_N\}$.

Definition (triangulation): A finite system of subsets $\mathcal{T} = \{T_1, ..., T_m; T_i \subset \Omega\}$ is called *triangulation* if all the T_i are pairwise disjoint open tetrahedrons whose closure covers Ω. The corners of the tetrahedrons are called *nodes*; the set of all nodes is denoted by $\mathcal{N}$. The subset $\mathcal{N}_{int} \subset \mathcal{N}$ denotes the set of *inner nodes*, the nodes that do not lie on $\partial \Omega_h$.

1. Finite elements

In other words, a triangulation is a decomposition of the domain into triangles/tetrahedrons where different triangles/tetrahedrons do not overlap but every point of Ω lies on one such subdomain or along the interface of two of them. It should be pointed out that a perfect covering of Ω can only be ensured if $\partial\Omega$ is piecewise linear. We will not stress the point of boundary approximation by domain triangulation here, for further reading see e.g. Oden, 1976. For the construction of an approximate vector space V_h not all triangulations are appropriate. Instead only the so called *regular triangulations* are suitable (Ciarlet, 1978):

Definition (regular triangulation): A triangulation $\mathcal{T}$ is *regular* if for $\overline{T}_i \cap \overline{T}_j, i \neq j$, exactly one of the following properties holds:
a) $\overline{T}_i \cap \overline{T}_j = \emptyset$
b) $\overline{T}_i \cap \overline{T}_j$ is a node of T_i and T_j
c) $\overline{T}_i \cap \overline{T}_j$ is an edge of T_i and T_j

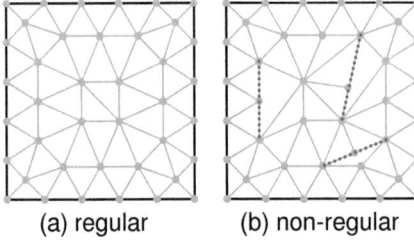

(a) regular (b) non-regular

Figure 1.3: Different domain triangulations of a polygonal domain. (a) shows a regular triangulation; the intersection of neighbouring elements are either a node or a whole edge. (b) represents a non-regular triangulation; the highlighted edges show deviations.

This definition basically says that different triangles adjacent to each other always have either nodes or *whole* edges in common. Fig. 1.3(a) presents a possible regular triangulation while Fig. 1.3(b) shows a non-regular one; the triangles not satisfying the definition above are highlighted.

Based on a regular domain triangulation, a set of basis functions can be constructed. As an example, Figure 1.4 shows a schematic representation of a so-called *hat function*. The basis function $\Lambda_i : \Omega \to \mathbb{R}$, $i = 1,...,|\mathcal{N}|$ is piecewise linear on every triangle, i.e. for every $k \in \{1,...,|\mathcal{T}|\}$ it may be written in the following form:

$$\Lambda_i(x) = a_{ik} + \langle b_{ik}, x \rangle, \qquad a_{ik} \in \mathbb{R}, b_{ik} \in \mathbb{R}^n \qquad \forall x \in T_k$$

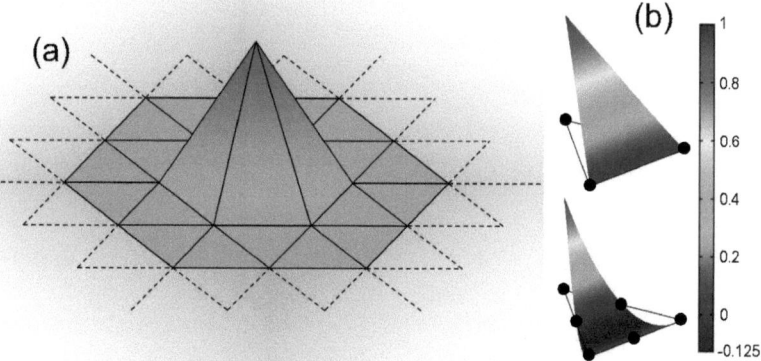

Figure 1.4: Examples of Lagrangian basis functions. (a) A given basis element obtains the value "1" at a single node point and is "0" at all other nodes. (b) Lagrangian elements of order "1" (upper plot) and "2" (lower plot). In the case of orders ≥ 2 basis functions are not completely specified by their values at the nodes anymore; additional points need to be chosen. For 2nd order Lagrangian elements, the centres of each edge are chosen (black markers).

Figure 1.4 shows a special case, the so-called *Lagrange-elements*. Each basis function has the value "1" at a specified node and "0" at all other nodes. Denoting the node positions by $\{x_j\}_{j=1,...,|\mathcal{N}|}$, the basis functions additionally satisfy $\Lambda_i(x_j) = \delta_{ij}$ with $i, j = 1,...,|\mathcal{N}|$. In general, *Lagrange-elements* of order k are given by polynomials of degree $\leq k$, examples are shown in the insets of Figure 1.4. Except for convection dominated advection-diffusion problems (compare section 2.1.8), such functions are employed as a basis in all the applications discussed in this work. With these preparations we are now able to define the finite elements:

Definition (finite elements): The Lagrangian hat functions $\{\Lambda_1,...,\Lambda_N\}$ in respect to a triangulation T form the *nodal basis*. The linear space $V^T = \text{span}\{\Lambda_1,...,\Lambda_N\}$ is the space of continuous, piecewise linear functions on Ω in respect to the triangulation T. The *finite elements* of Ω in respect to the triangulation T is defined by the tuple (T, V^T).

The following two theorems prove that the above-described construction of the approximate linear space is well suited for the problem at hand:

Theorem: The space V^T is a finite-dimensional subspace of $H^1(\Omega)$.

1. Finite elements

Theorem: Let $\mathcal{T}$ be a regular triangulation of Ω with nodes at $x_i, i=1,...,|\mathcal{N}|$. It exists a unique function $\psi \in V^{\mathcal{T}}$ that solves for given $y_i \in \mathbb{R}, i=1,...,|\mathcal{N}|$ the following interpolation problem

$$\psi(x_i) = y_i \quad i=1,...,|\mathcal{N}| \qquad \text{given by} \qquad \psi(x) = \sum_{i=1}^{|\mathcal{N}|} y_i \Lambda_i(x)$$

To validate this approach, it is necessary to know how far away the approximate solution u_h is from the exact one u. For elliptic equations, it is possible to show that the error $\| u - u_h \|_{H^1(\Omega)}$ can be estimated by

$$\| u - u_h \|_{H^1(\Omega)} \leq C \inf_{v \in V_h} \| u - v \|_{H^1(\Omega)}$$

with a constant C obtained from the *Poincaré inequality* (Triebel, 1980). In principle, this ensures that the solution is as accurate as it can be under the approximation V_h of the solution space.

1.2.4 Assembly and stiffness matrix

The main issues for the implementation of finite element methods are the creation or *assembly* of the stiffness matrix A introduced in (1.11) and the solving of the linear system. In regards to further applications, we will therefore briefly summarize some important facts. According to (1.11) the matrix elements are given by the integral expressions

$$A_{ij} = a(\Lambda_i, \Lambda_j) = \int_{\Omega} \langle \nabla \Lambda_i, \nabla \Lambda_j \rangle dx$$

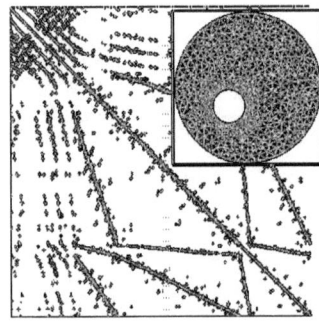

Figure 1.5: Domain triangulation and stiffness matrix of a finite element discretization; only the black entries are different from zero.

Since the hat functions Λ_i are only different from zero among a small number of elements, the integral is zero in most cases: in principle, a non-zero value is only obtained when hat functions to adjacent elements are considered. An example is shown in Figure 1.5. The geometry is discretized by a mesh consisting of 794 elements which leads to 1660 nodes. Therefore, the stiffness matrix has $1660^2 = 2755600$ entries. Figure 1.5 shows a plot marking all the non-zero values for the system. We find that only a small fraction of all entries, 18550, has non-zero values. Such a matrix is called *sparse*. Sparsity of a matrix allows for very memory efficient ways in which to handle it. In praxis, only the non-zero entries under the information of their matrix indices are saved; this is called the *assembly of the stiffness matrix*.

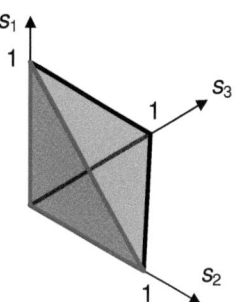

Figure 1.6: Representation of 2-(dark) and 3-(bright)simplex.

The triangles of the triangulation are usually of different forms. Therefore, instead of evaluating the actual form of all basis functions, the integral is recast onto a reference element, the unit simplices S_n given by

$$S_n = \{ s \in \mathbb{R}^n, s_i > 0, \sum s_i < 1 \} \quad (1.12)$$

Examples of 2- and 3-simplex are shown in Figure 1.6.

The transformation from an arbitrarily shaped triangle to the simplex is given by an affine mapping $\Phi_k : S_n \to T_k$. Considering the two-dimensional case and denoting the nodes of an element T_k by $x_{k,j}, j = 1,..,3$, Φ_k may be written in the explicit form

$$\Phi_k(s_1, s_2) = x_{k,1} + s_1(x_{k,2} - x_{k,1}) + s_2(x_{k,3} - x_{k,1}) \quad (1.13)$$

1.2.5 Parabolic equations and time integration

The systems considered in the preliminary sections describe a stationary problem; the resulting solutions do not show an evolution in respect to time but may be seen as a solution for the long time limit $t \to \infty$. If we are interested instead in the relaxation of a disturbed state back to equilibrium or simply need to handle data changing over time, an additional time argument must be introduced. Let be $I = [0;T] \subset \mathbb{R}$ a finite time interval, we address the following problem:

1. Finite elements

> find $u: I \times \Omega \to \mathbb{R}$, such that
> $$\partial_t u - \Delta u = f(x,t) \quad \text{for } x \in \Omega, t \geq 0$$
> $$\langle \hat{n}, \nabla u(x,t) \rangle = 0 \quad \text{for } x \in \partial\Omega, t \geq 0 \qquad (1.14)$$
> $$u(x,0) = u_0(x) \quad \text{for } x \in \Omega$$

From a formal point of view, we went over from the theory of elliptical equations to parabolic systems. Writing the coefficients of all derivatives of order 2 in a matrix B, all eigenvalues λ except of one are either $\lambda > 0$ or $\lambda < 0$. The remaining value is given by $\lambda = 0$. For the complete specification of the problem, an additional initial state needs to be specified which is given by the initial data $u_0 : \Omega \to \mathbb{R}$.

For the treatment of such systems, it is no longer sufficient to merely decompose the domain within a Galerkin approach; the time argument also needs to be discretized. To do so, we reformulate (1.14) into its corresponding weak form (in a manner similar to the discussions in section 1.2.1)

$$\int_\Omega \frac{\partial u(t)}{\partial t} \psi \, dx + \int_\Omega \langle \nabla u(t), \nabla \psi \rangle \, dx = \int_\Omega f(t) \psi \, dx \qquad \forall \psi \in H^1(\Omega)$$

or
$$\left\langle \frac{\partial u(t)}{\partial t}, \psi \right\rangle_{L^2(\Omega)} + a(u(t), \psi) = \langle f(t), \psi \rangle_{L^2(\Omega)} \qquad (1.15)$$

We declare a weak solution of (1.14) as a function $u : \Omega \times I \to \mathbb{R}$ satisfying (1.15) and passing to the limit $\|u(t) - u_0\|_{L^2(\Omega)} \to 0$ for $t \to 0$. Further, we demand $u(t) \in H^1(\Omega)$ for almost every $t > 0$. It is possible to show the existence of a weak solution for arbitrary $f \in L^2(\Omega \times I)$ and $u_0 \in H^1(\Omega)$ where $u(t, \cdot) \in H^1(\Omega)$ for every $t \geq 0$. Additionally, for almost every $t > 0$, it is $u(t, \cdot) \in H^2(\Omega)$ and $\partial_t u(t, \cdot) \in L^2(\Omega)$. For u to be also a strong solution requires f to be twice continuously differentiable and the initial conditions to maintain the compatibility

$$\langle \hat{n}, u_0(x) \rangle = 0 \quad \text{and} \quad -\Delta u_0(x) = f(x,0) \text{ for } x \in \partial\Omega$$

In respect to the spatial argument parabolic equations are very close to the initially discussed elliptic ones. Therefore, the idea is close to apply a Galerkin discretization in respect to space. Analogously, we may construct a finite-dimensional linear space

1.2 Introduction

$V_h \subset H^1(\Omega)$ from which the approximated solutions need to be chosen. Addressing the problem

find $u_h \in V_h$, such that

$$\left\langle \frac{\partial u_h(t)}{\partial t}, \psi \right\rangle_{L^2(\Omega)} + a(u_h(t), \psi) = \langle f(t), \psi \rangle_{L^2(\Omega)} \quad (1.16)$$

$$\forall \psi \in V_h$$

This way, an approximated problem is received which is discretized in respect to space but continuous in regards to the time argument: a pre-step towards the completely discrete model which is known as *semidiscretization*. An approximate solution may be written in respect to the nodal basis $\{\Lambda_1, ..., \Lambda_N\}$ via the expansion:

$$u_h = \sum_{j=1}^{N} u_j(t) \Lambda_j \quad (1.17)$$

with time-dependent coefficients $u_j : I \to \mathbb{R}$; the hat functions are still independent of the time argument. Inserting (1.17) into equation (1.16) and exploiting the bilinearity of the scalar product, we may write:

$$\sum_{j=1}^{N} \frac{\partial u_j(t)}{\partial t} \langle \Lambda_j, \Lambda_i \rangle_{L^2(\Omega)} + \sum_{j=1}^{N} u_j(t) \cdot a(\Lambda_j, \Lambda_i) = \langle f(t), \Lambda_i \rangle_{L^2(\Omega)}$$

Again, we obtain a matrix formulation by introducing the vectors $U, b \in \mathbb{R}^N$ by $U_j(t) = u_j(t)$ and $b_j(t) = \langle f(t), \Lambda_j \rangle_{L^2(\Omega)}$:

$$M \frac{\partial U}{\partial t} + AU = b \quad (1.18)$$

with $M_{ij} = \langle \Lambda_i, \Lambda_j \rangle_{L^2(\Omega)}$ and $A_{ij} = a(\Lambda_i, \Lambda_j)$

M is called the *Gramian matrix* of the hat functions in $L^2(\Omega)$. The theorem of Picard-Lindelöf (see e.g. Hackbusch, 1996) ensures the existence and uniqueness of a solution U which remains true also for quasilinear problems, i.e. where the matrix M is linear in U (Nobile, 2001). Due to graphical reasons, such a semidiscretization is

1. Finite elements

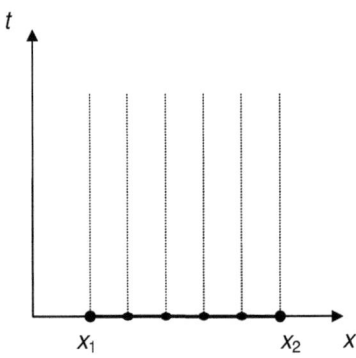

Figure 1.7: Method of lines for the one-dimensional case $\Omega = [x_1, x_2]$. Discrete domain points evolve along a continuous time argument.

also called *(vertical) method of lines*: for every $t \geq 0$ the vector valued function $U(t)$ evolves at discrete domain points along a continuous time argument as schematically shown in Figure 1.7. In a second step, a discretization of the time argument is necessary. Throughout this work, a backward differential formula (BDF) of variable order ≤ 5 is used for numerical time integration schemes. Their analysis is part of the theory of numerical treatment of ordinary differential equations, we will therefore not stress this point here any further but refer to standard literature, e.g. Butcher, 1987 or Ascher and Petzold, 1998.

1.3 Living on a bubble – Moving domains

In section 1.2.5, we explained how additional time arguments need to be incorporated into the numerical scheme. However, the discussion above does not include a very important case which is given by systems where the domain Ω itself evolves along the time interval I. In real life such changes of the domain are often encountered when analysing physical systems where a liquid/gas flow interacts with a solid material. Typical examples can be found e.g. in haemodynamics where blood vessels expand and contract during the heartbeat due to forces induced by the blood flow; in the analysis of buildings such as skyscrapers which bend in the wind; or when an air bubble rises in liquid. As long as such deformations are small, they can be treated in the framework of linear elasticity theory; the change of the domain shape can be incorporated by additional boundary conditions which are also called *transpiration conditions*. This is no longer possible if large displacements occur and the coupling between different physical effects obtains an increasingly non-linear character. We will give two examples of the possible different approaches dealing with strong deformations:

(a) the Level-set-method
(b) the ALE-method

1.3.1 Level-set-method

The idea of the Level-set-method is to introduce a continuous function $\Phi : \Omega \to \mathbb{R}$. Due to continuity the sign-function decomposes the image $\Phi(\Omega)$ into different sections which may be used to model different materials. Therefore, the implicit function g defined by $\Phi(g) = 0$ describes material interfaces. A typical example is shown in Figure 1.8: A gas bubble rises within a liquid (top pictures), the initial configuration of Φ is shown on the left side. The lower plots show a domain decomposition according to the sign-function; here blue and red areas correspond to $\Phi(x) < 0$ and $\Phi(x) > 0$, respectively.

The evolution of the function Φ is described by a convection-diffusion equation which will be discussed in more detail in section 2.1.6. For right now, we just accept that it is given by the parabolic equation

$$\frac{\partial \Phi}{\partial t} - D\Delta\Phi + \nabla(u\Phi) = R \tag{1.19}$$

The dynamic change of Φ is due to a convective transport of Φ along a flow u and a reaction rate R. In principle, these contributions are sufficient. However, if we take a closer look at the lower plots of Figure 1.8, we see that fine structures are generated

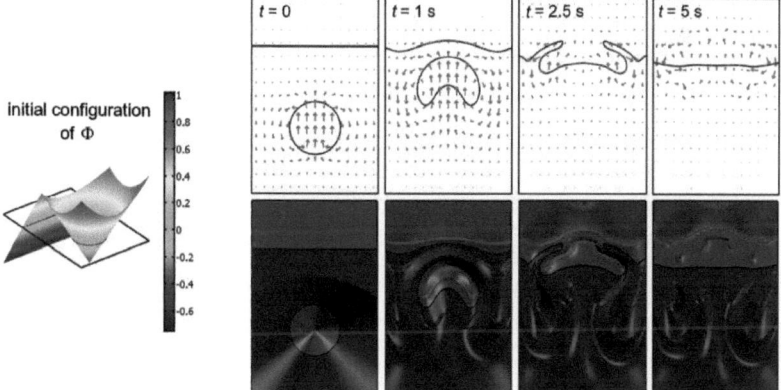

Figure 1.8: The rising bubble; example of a moving domain calculated in a Level-set-framework. The left plot shows the initial configuration of the function Φ. The right plots show the implicit function g given by $\Phi(g) = 0$ indicating fluid-gas-interfaces (black lines) that are moved along the flow profile (arrows). Due to the inhomogeneous velocity distribution, the bubble surface undergoes deformation.

1. Finite elements

close to the interfaces. The reason for such behaviour will be explained in 2.1.7. Such details start appearing when the convective flow dominates the diffusive one, i.e. $|D\nabla\Phi| \ll |u|$. Higher diffusion ensures convergence of the numerical scheme; the term $-D\Delta\Phi$ therefore maintains numerical stability. The choice of D itself is very important, as here it has no physical interpretation but is only introduced for technical reasons. On the one hand it needs to be large enough to average out numerical oscillations. On the other hand, it needs to be sufficiently small to not introduce non-physical results.

A good introduction into the basic ideas and the formalism can be found in Osher and Ronald, 2003. Chang et al., 1996, give a good example how this method can be applied to fluid flow calculations.

1.3.2 ALE-method

One of the main disadvantages of Level-set-method based approaches is the fact that in most cases a bigger domain than the one of interest needs to be considered. The reason for this is that the domain travels through a fixed space, one which must include *all* possible positions of the object studied. The description matches the point of view of a stationary observer; the corresponding coordinate frame is called *Eulerian* or *spatial frame*. In continuum mechanics, a different coordinate system is commonly employed, the *Lagrangian frame*. In contrast to the spatial description, the point of view chosen is that of an observer travelling with the material. In his frame, the observer remains at the same position all of the time; therefore, the domain remains non-moving and can be written in the form of a *reference configuration* Ω_0.

To avoid confusion, we will denote the domain configuration in the spatial frame at time t by Ω_t; coordinates in this frame by x; while ξ denotes the coordinates in the Lagrangian or reference frame. The idea of the ALE-method (*Arbitrary Lagrangian Eulerian*) is to transfer the physical phenomena described in the spatial system onto the reference frame and therefore onto a non-moving domain which enables the application of the methods discussed in section 1.2. For this purpose, we introduce a family of mappings $\mathcal{A}_t : \Omega_0 \to \Omega_t$ holding several properties:

a) $\mathcal{A}_t : \Omega_0 \to \Omega_t$ is a homomorphism: for every $t \in I$, i.e. $\mathcal{A}_t \in \mathcal{C}(\bar{\Omega}_0)$ and $\mathcal{A}_t^{-1} \in \mathcal{C}(\bar{\Omega}_t)$

b) the application $t \mapsto \mathcal{A}_t(\xi, t)$ is differentiable almost everywhere in I

1.3 Moving domains

The connection between the reference system and the spatial configuration via the *ALE-mapping* $\mathcal{A}_t$ is schematically shown in Figure 1.9. To get a better understanding of how the spatial equation can be recast into the reference frame, we will address the parabolic equation (1.14) again.

A function f declared on the spatial coordinate frame induces a function $\tilde{f}$ on the Lagrangian frame given by $\tilde{f}(\xi,t) = f(\mathcal{A}_t(\xi),t)$. Time derivatives ∂_t of f and $\tilde{f}$ are to be understood in respect to the corresponding fixed coordinate frame. To transfer between both frames, another definition of time derivative is often necessary: the *material derivative*. This derivative may be visualized as the time derivative of a function f along the trajectory of a reference coordinate ξ (compare Figure 1.9). In formula, it is

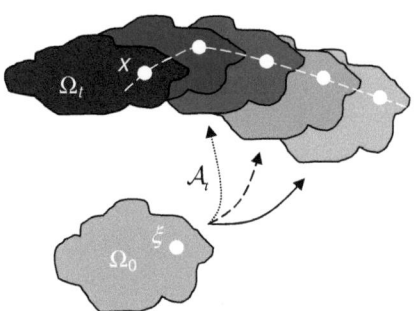

Figure 1.9: Mapping of reference configuration Ω_0 in the Lagrangian frame onto the spatial configuration Ω_t.

$$\frac{Df}{Dt}(x,t) := \frac{d}{dt} f(\mathcal{A}_t(\xi),t) = \sum_{i=1}^{n} \frac{\partial f(\mathcal{A}_t(\xi),t)}{\partial x_i} \frac{d\mathcal{A}_t(\xi)_i}{dt} + \frac{\partial}{\partial t} f(\mathcal{A}_t(\xi),t)$$

$$= \left\langle \nabla f(x,t), \frac{d\mathcal{A}_t(\xi)}{dt} \right\rangle + \frac{\partial f}{\partial t}(x,t)$$

$$= \langle \nabla f(x,t), w(x,t) \rangle + \frac{\partial f}{\partial t}(x,t)$$

with $\quad w(x,t) = \dfrac{d\mathcal{A}_t(\xi)}{dt} \quad$ (1.20)

with w the *domain velocity*. Again, a visualization of each component is possible: an observable f evaluated along the trajectory of a reference point ξ in the spatial system changes in respect to time due to the motion of the coordinate ξ with velocity w (1st term) as well as the temporal evolution of f itself (2nd term). This is schematically shown in Figure 1.10.

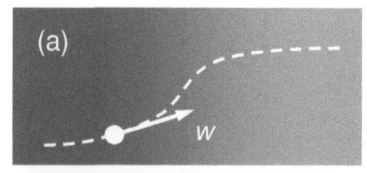
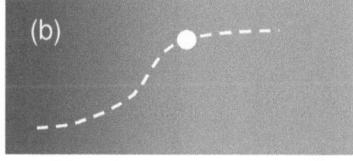

Figure 1.10: Change of a function f along a trajectory in respect to time. The value change is due the change by moving with velocity w (a) and by the time dependence of f itself (b).

1. Finite elements

With the help of the ALE-mapping, it is possible to recast equations formulated in the spatial system onto the reference configuration. Considering again the parabolic equation (1.14), the weak formulation (1.16) goes over to

$$\left\langle \frac{\partial u(t)}{\partial t}, \psi \right\rangle_{L^2(\Omega_t)} + a_t(u(t),\psi) = \langle f(t),\psi \rangle_{L^2(\Omega_t)}$$

with $\quad a_t(u(t),\psi) = \int_{\Omega_t} \langle \nabla u(t), \nabla \psi \rangle dx$

for every $\psi \in V_h$, where the integrations need to be extended to the time depending domains Ω_t. According to the substitution formula (compare e.g. Forster, 1999) an integration of a function $f : \Omega_t \to \mathbb{R}$, $f \in L^1(\Omega_t)$ can be recast to the domain Ω_0 via a mapping $\mathcal{A}_t : \Omega_0 \to \Omega_t$ by

$$\int_{\Omega_t} f(x)dx = \int_{\Omega_0} f(\mathcal{A}_t(\xi)) \cdot |\det J_{\mathcal{A}_t}(\xi)| d\xi$$

with $J_{\mathcal{A}_t}$ the Jacobian of $\mathcal{A}_t$. Here, we may therefore rewrite the parabolic equation into

$$\int_{\Omega_0} \tilde{\psi}(\xi) \frac{\partial (\tilde{u} \det J_{\mathcal{A}_t})}{\partial t}(\xi,t) d\xi$$

$$- \int_{\Omega_0} \tilde{\psi}(\xi) \det J_{\mathcal{A}_t}(\xi,t) \cdot \sum_{i,j} \frac{\partial \mathcal{A}_{t,j}^{-1}}{\partial x_i}(x,t)\bigg|_{x=\mathcal{A}_t(\xi)} \cdot \frac{\partial (\tilde{w}\tilde{u})_i}{\partial \xi_j} d\xi$$

$$- \int_{\Omega_0} \det J_{\mathcal{A}_t}(\xi,t) \cdot \sum_{i,j} \frac{\partial \tilde{u}}{\partial \xi_i} \frac{\partial \mathcal{A}_{t,j}^{-1}}{\partial x_i}(x,t)\bigg|_{x=\mathcal{A}_t(\xi)} \cdot \frac{\partial \tilde{\psi}(\xi)}{\partial \xi_j} d\xi$$

$$= \int_{\Omega_0} \det J_{\mathcal{A}_t}(\xi,t) \cdot \tilde{\psi}(\xi) \tilde{f}(\xi,t) d\xi \qquad \forall \tilde{\psi} \in \mathcal{Y}(\Omega_0) \qquad (1.21)$$

The testfunctions $\tilde{\psi} \in \mathcal{Y}(\Omega_0)$ are commonly formulated in respect to the reference configuration. Their choice is connected to the space in which the ALE-mapping is discretized and needs to maintain the suitability of the finite element discretization (Nobile, 2001). The obtained formulation is far more complex, however, it is applicable to a wide range of systems (an exception are contact phenomena, see below).

1.3 Moving domains

Though the application of ALE-based methods has gained high popularity, a rigorous mathematical analysis is still missing. In this regard though, the *geometrical conservation laws* have drawn a lot interest during the last decade. We do not want to go into detail here, but only remark that a numerical scheme meets the geometrical conservation laws if certain geometrical properties are maintained in respect to time by the numerical scheme. For finite volume schemes the works of Lesoinne and Farhat, 1997, and Guillard and Farhat, 2000, have identified these laws as a minimal condition on the precision of the quadrature formula used for numerical time integration. Similar observations have been made for finite element schemes, e.g. Masud, 2006, or Boffi and Gastaldi, 2004. However, a proper clear-cut analysis is not available so far.

Figure 1.11: Contact of two spheres modelled in an ALE-framework. When the spheres get close to each other, the mesh quality drops rapidly and mesh degenerates.

A disadvantage of schemes based on ALE-approaches is that they are not adaptable to systems in which topological changes occur; in particular, it is not possible to model contact phenomena in such a framework. The reason lies in the continuity requirement of the ALE-mapping which does not allow mapping two points separated in the reference frame onto the same coordinate in the spatial system. From a practical point of view, in many cases problems quickly arise if overly strong deformations are considered (compare Figure 1.11).

1. Finite elements

Chapter 2

Particles in microfluidic devices

Since the early 1990s, microfluidic lab-on-a-chip systems have been the focus of much research. One of the most challenging tasks is the construction of a *Micro Total Analysis System* (μTAS), the integration of several laboratory procedures on a small microfluidic chip (Pamme, 2006b; Gijs, 2004; Whitesides, 2006). These tasks include the *injection* and the *preparation* of the sample and the subsequent guidance by e.g. hydrodynamic or electromagnetic means to the functional sites of the device. It is at these sites where chemical *reactions* take place, followed by *separation* and *detection* of the reaction products. Each individual component produces a unique set of challenges which have been thoroughly studied during the last decade. Due to the considered size scales, the fluid flow itself does not behave in a way how macroscopic observations would predict. Inertia effects and non-linear dynamics play only a minor role on the investigated scales (for special exceptions see e.g. Carlo, 2009; Blattert, 2005). In particular, this results in severe complications when trying to mix several fluid components on the microscale with each other. Where it is sufficient to pure two components together on the macroscale, a lot of effort is necessary (see e.g. Kamio et al., 2009; Niu et al., 2006; Long et al., 2009; or Lee et al., 2009) to reach a similar result when dealing with microfluidic devices.

μTAS-devices require for the controlling of the component that is to be analyzed. Manipulation techniques of objects dissolved in liquid by applying external influence include electrophoretic (Green et al., 1998), dielectrophoretic (Green et al., 2000;

2. Microfluidic devices

Yang and Lei, 2006), electrothermal (Green et al., 2001) and magnetophoretic effects (Häfeli et al., 2005; Pamme and Wilhelm, 2006c). What effect can be used for a specific task depends mainly on the properties of the component itself, e.g. it needs to be magnetic to feel a force in an inhomogeneous magnetic field and therefore enable magnetophoretic procedures. The prototype of magnetic components is a magnetic micro- or nanoparticle. During the last years, a lot of effort has been done for the creation of particle with different properties (Hütten et al., 2005; Sun, 2006) and their surface modifications (Woo et al., 2005; Mornet et al., 2005). In particular, FePt-particles (Chen et al., 2006) have recently attracted a lot of interest due to their strong anisotropy which allows for their application in data storage devices (Moser et al., 2002).

Many strategies for their manipulation have been developed in recent years. In particular, current leading wires which create strongly inhomogeneous magnetic fields have been applied in many different ways (Wang et al., 2006). Another method is the employment of magnetic components. I. Ennen et al., 2007, observed the self-assembly of magnetic nanoparticles along the domain walls in patterned ferromagnetic layers, whereas T. Deng et al., 2001; E. Mirowski et al., 2007, and G. Vieira et al., 2009, demonstrated a particle transport via magnetic micro-components which switched their magnetic configuration periodically in respect to time. Their easy handling enables a lot of different applications such as e.g. their purification in respect to certain properties (Afshar et al., 2009). By tagging them to biomolecules such molecules are indirectly accessible to magnetophoretic manipulation themselves. In particular, we may guide biomolecules through a μTAS-device employing magnetic fields. This strategy is pursued in this work.

The components of the microfluidic device discussed in this work are schematically shown in Figure 2.1. Assuming blood as a carrier liquid, the sample is inserted into the lab-on-a-chip structure via a microfilter that separates the blood plasma from the serum, which is necessary to prevent the closing of the channel by aggregation of colloidal components within the plasma (i.e. clotting). The sample is guided by microfluidic channels to a *preparation/reaction* chamber (Figure 2.1(b)). In our case, the biological binding of biofunctionalized magnetic markers to the antigens of interest would take place in this site. In general, this type of chamber can be used for all kinds of reaction applications. Upon leaving the reaction site, the solution is transported to a *separation* device (Figure 2.1(c)) where different surfactants are purified from each other and travel to a detection site (Figure. 2.1(d)). Due to their magnetic, properties small magnetic particles influence soft magnetic material nearby, which creates the possibility to detect them with magnetoresistive sensors. If the surface of the sensor itself is specifically coated, the corresponding antigen can act as a linker between the sensor and magnetic marker. Depending on the response of the sensor, it is therefore possible to indirectly detect the presence of a certain antibody and thus to diagnose a certain infection.

Individual components have different requirements for the transport properties; therefore, one of the most challenging aspects is the combination of all these tasks. Along the reaction site, certain time scales must be maintained in order to enable chemical reactions and biological binding processes. Furthermore, the transport should show a diffusive character rather than a convective one, as this facilitates high reaction probability. This changes for the separation site, where instead high stability of the flow is necessary to maintain a degree of separation purification. This is required in order to make proper estimations according to the signal measured at different detection devices. Therefore, different components must be designed not only to handle tasks, but also to be compatible with one another. Simplicity is also highly desirable when adapting individual components from the laboratory stage to the on-

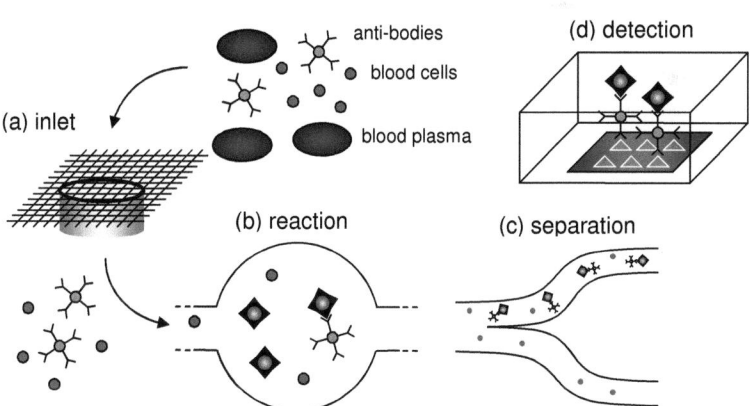

Figure 2.1: Micro Total Analysis System (for haematological applications) (a) The sample enters the microfluidic structure via a microfilter which separates big colloidal components and (b) reaches a preparation / reaction site. (c) Depending on their functionalization, a different behaviour along the separation device can be used to guide components to a specific detection site (d) enabling the estimation of whether a certain component was present in the original sample or if a certain reaction has taken place.

2. Microfluidic devices

chip structure. In particular, this includes avoid electric components (e.g. current leading wire geometries) as much as possible. Therefore, the goal of this chapter is to introduce solutions for different tasks that are as simple as possible. The proposed designs may also help to reduce the complexity of existing lab-on-a-chip devices.

First of all, when analyzing the behaviour of magnetic particles dissolved in a moving fluid, it is necessary to understand the dynamics of the liquid itself. In section 2.1, we will therefore give a brief review of the fundamental ideas and equations of hydrodynamics, focusing in particular on the special case of microfluidics. Sections 2.2 to 2.4 will then discuss different parts of the microfluidic chip. The separation device (chapter 2.2) is published in a work entitled *A hydrodynamic switch: A separation device for magnetic beads*, Applied Physics Letters **95**, 2009, while the results presented in chapter 2.3 can be found in *A combined reaction-separation lab-on-a-chip device for low Péclet number applications*, Journal of Applied Physics **106** (2), 2009. The positioning structure discussed in section 2.4 was recently released under the title *Positioning system for particles in microfluidic structures*, Microfluidics and Nanofluidics **7** (6), 2009.

The detection mechanism itself will be discussed in detail in chapter 5. For the design of the microfluidic guidance structure, it is only necessary to know that a proper detection can only be ensured if particles are close enough to the sensors i.e. the bottom of the channel. In this entire chapter, we will consider a very low particle concentration which allows for the assumption of neglectable particle-particle interactions. For high concentrations, different effects can be observed which is discussed in the sections 4.5 and 4.6.1.

2.1 Fundamentals of hydrodynamics

Hydrodynamics deal with the description of fluids which can be either gases or liquids. In this work, we will focus on the latter case. On the microscopic level, such liquids consist of a large amount of particles $\sim \mathcal{O}(10^{23})$ interacting with each other. In principle, to achieve an exact description of the state of the liquid at every time, a full knowledge of the position and momentum of each individual component would be necessary. However, this problem cannot be handled due to the resulting number of degrees of freedom; therefore, an effective theory must be applied. The basic princi-

2.1 Fundamentals

ples of how to bridge the gap between the discrete microscopic description and the continuous mesoscopic level will be explained in this chapter.

2.1.1 Continuum hypothesis and effective parameters

It is not possible to completely analyze the motion of each fluid particle at a given time. (Even if there would be a way to obtain such information, the result would be of little help as we do not observe the actual microscopic processes in experiments on the macroscale but instead witness an effective dynamic resulting as some sort of average.) In this sense, an effective theory enabling direct access to a small set of dependent variables would be not only much easier to handle from a theoretical point of view, but also much more suitable for the actual description of the system. A basic requirement of an effective theory is the possibility to average the microscopic details in respect to space as well as time. In the framework of hydrodynamics, two assumptions, known as the *continuum hypothesis*, must be met. If this requirement is fulfilled, it is possible to approach the discrete liquid by a continuum model.

1) A separation of length scales is possible

Let A be an observable evaluated at a certain room point r. If we consider instead of the point r itself a small cube of volume dV around r and calculate the average $\langle A \rangle_{dV}$ of A along dV, the obtained value depends on the volume of dV as shown in Figure 2.2. For small elements, at the size scale of the intermolecular distance (~ 0.3 nm for liquids and 3 nm for gases, Bruus, 2008) strong fluctuations can be found due to the discrete structure of the fluid. If we enlarge the volume and therefore average A over a higher number of microscopic details, the *mesoscopic area* is obtained. Fluctuations even out and the value of $\langle A \rangle_{dV}$ is independent of the exact size of dV. If

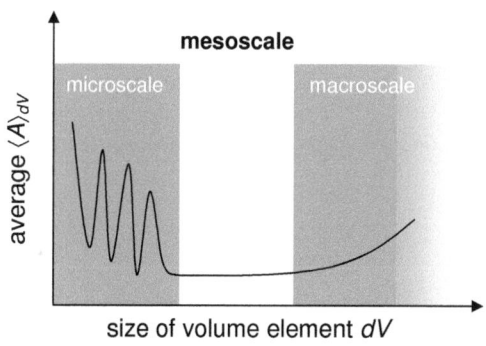

Figure 2.2: Dependency of $\langle A \rangle_{dV}$ on the volume size dV. On the microscale strong fluctuations can be found that average out for larger element sizes.

2. Microfluidic devices

volume sizes on the macroscale are considered, the average is taken over the regime that is supposed to be studied. Even external influences may vary on this scale.

We may approximate the small volume dV by a cubic element of side length λ. By choosing a value $\lambda \sim 10$ nm, we can reasonably expect to obtain mesoscopic behaviour. The resulting cube contains $\sim \mathcal{O}(10^4)$ particles and is therefore large from a microscopic point of view, but small from a macroscopic one. The small elements of volume dV will be called *volume elements* in the following. If we talk about a certain property of a liquid at point r, we actually mean the average of that property along the volume element situated at r.

2) A separation of time scales is possible
We will further assume that along every small volume element dV the system has reached thermal equilibrium in respect to a strongly reduced set of (space-dependent) variables; such variables are referred to as *slow variables*. Good candidates for these are densities of conserved quantities e.g. mass, momentum or energy density.

In this approach, all micro processes result in an effective behaviour of the small liquid element which can be completely described by two fields: a vector field $u: \mathbb{R}^n \times \mathbb{R} \to \mathbb{R}^n$, describing the *velocity* of a volume element at each space-time-point, and a scalar field, denoting the pressure $p: \mathbb{R}^n \times \mathbb{R} \to \mathbb{R}$ within the liquid. By going from the micro- to the mesoscale, different phenomena cannot be described any longer. For example, the interaction between single molecules is "lost" by averaging along volumes, which always contain a large number of degrees of freedom. Therefore, the model includes certain material parameters that cannot be calculated in this framework but need to be obtained by either experimental observations or models on the microscale. Three different parameters are important for our analysis: the *compressibility* κ, the *density* ρ and the *viscosity* η.

Compressibility is defined by

$$\kappa = -\frac{1}{V}\frac{\partial V}{\partial p} \qquad (2.1a)$$

and is a measure for the relative volume change in respect to pressure. Typical values can be found in table 2.1; it should be pointed out that values for liquids are much smaller than those for gases. Indeed, compressibility effects of liquids are commonly small enough to be neglected. Therefore, we will assume $\kappa = 0$ in the following which describes the special case of *incompressible liquids*.

The density ρ is the ratio between the mass and the volume of a liquid element at a point $\mathbf{r}$. Since the fluids regarded are incompressible, the ratio between mass and volume is a constant along the whole volume. Thus, denoting the total mass and volume by an index *tot*, it is

$$\rho = \frac{m_{tot}}{V_{tot}} \qquad (2.1b)$$

The viscosity η is a measure for the inner friction of the fluid. Experimentally, it can be determined by the setup shown in Figure 2.3: two plates are separated by an initially non-moving liquid film; the upper plate is free while the lower is chosen fixed. Exerting a force $\mathbf{F}$ parallel to the plane direction induces a *shear stress* $\tau = F/A$ with A the surface area of the plane. This shear stress acts onto the liquid and induces a fluid flow. Due to the symmetry of the system, the resulting velocity is parallel to the applied force and only depends on the y-direction, $\mathbf{u} = u_x(y)\hat{\mathbf{x}}$. The derivative $\partial_y u_x(y)$ is called *strain rate* and is obviously connected to the applied shear stress $\tau = f(\partial_y u_x(y))$. In the special case of a linear relation

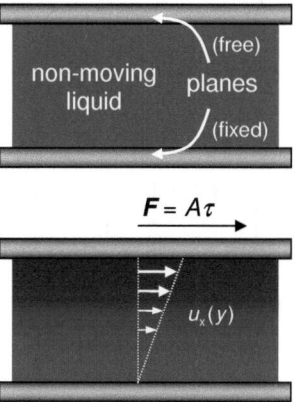

Figure 2.3: Experimental setup for rheology measurements. A fluid flow is induced by shear stress.

$$\tau = \eta \frac{\partial u_x(y)}{\partial y} \qquad (2.1c)$$

the fluid is called *Newtonian fluid* and the parameter η stands for the *viscosity* of the liquid. In the present work, we will restrict our analysis to Newtonian fluids covering the case of water which was used in all the experimental setups. Typical examples of non-Newtonian liquids are polymer solutions or blood but also many examples we find regularly in everyday life such as paint, shampoo or ketchup. The viscosity of

2. Microfluidic devices

ketchup decreases under high shear rate; therefore, it needs to be shaken before it can be poured out of a bottle.

The introduced parameters are supposed to be independent of space or time. It can be implied from this assumption that all of our discussions must be based on a constant temperature T since all coefficients show slight temperature dependence. Therefore, a spatially varying temperature distribution would introduce an implicit spatial dependence of the material parameters. Depending on the temperature gradients found in such systems, different phenomena may occur. A well known example is the *Rayleigh-Bérnard convection* which can be found e.g. in lava lamps: the temperature along a certain volume fraction increases leading to a decreasing density ρ. If this change occurs sufficiently fast, it introduces a density gradient and the hot, lighter-volumed fractions ascend due to buoyancy. At a certain height, they cool down again and drift back down.

2.1.2 Lagrangian and Eulerian frame

For the analysis of hydrodynamic systems two different representations can be chosen which are known as *Lagrangian* and *Eulerian frame*. As already explained in section 1.3, the Eulerian frame is referred to as spatial representation and coincides with the view of a resting observer. The liquid flows in respect to a spatially fixed coordinate system: at an arbitrary space point different volume elements can be found at different times. This seems to be a natural way of how to describe certain phenomena. However, it does not go together with the description of a liquid introduced in the preliminary section which identifies small volume elements as the basic components and also of the "carrier" of physical properties.

In chapter 1.3, we already calculated how a variable A changes in respect to time within the two different frames. In particular, we introduced the *material derivative* D/Dt as the time derivative for a fixed reference coordinate along its trajectory within the spatial frame. In fluid dynamics, the volume elements correspond to the reference coordinates. Therefore, according to equation (1.20), the time evolution of the variable A at a fixed liquid element can be written as

$$\frac{DA}{Dt}(r,t) = \left\langle \frac{\partial A}{\partial r}(r,t), u \right\rangle + \frac{\partial A}{\partial t}(r,t)$$

with $\partial/\partial t$ the usual time derivative and u the velocity of the volume element, i.e. the flow velocity. The above formula can be readily extended to a vector-valued observable A by replacing the first summand on the right hand side by $J_A u$ with J_A the Jacobian of A. However, in general literature, the symbolic expression $(u\nabla)A$ is used throughout, we will therefore write

$$\frac{DA}{Dt} = (u\nabla)A + \frac{\partial A}{\partial t} \qquad (2.2)$$

2.1.3 Navier-Stokes equation and Reynolds number

The governing equations for velocity and pressure field can be derived from two general physical principles: mass and momentum conservation. As already mentioned, the density is considered to be constant along the whole liquid. A mass flow towards a certain space point r is therefore only possible if there is a mass flow of the same size away from r. In other words, the mass flow ρu does not have any sources or drains and therefore needs to satisfy the *equation of continuity*

$$\nabla(\rho u) = 0,$$

which in our case can be further simplified due to the assumption of incompressibility

$$\nabla u = 0. \qquad (2.3)$$

The motion state u of the liquid can be obtained from momentum conservation. The forces influencing the motion state originate from shear stresses and pressure gradients along the flow. In general, applying Newton's second law on a volume element dV with boundary Γ and surface normal vector $\hat{n}$, it is

$$\rho \int_{dV} \frac{Du}{Dt} dr = \int_{\Gamma} (-p \cdot \mathrm{Id} + \sigma) \hat{n} \, dr + \rho \int_{dV} f \, dr \qquad (2.4)$$

denoting by Id the identity matrix in n dimensions and by σ the stress tensor. Further, f refers to external force densities acting on the liquid which can arise from e.g. grav-

2. Microfluidic devices

ity or the coupling to electromagnetic fields. In the special case of incompressible Newtonian liquids, this equation may be reformulates as the *Navier-Stokes equation*

$$\frac{\partial u}{\partial t} + (u\nabla)u = -\frac{\nabla p}{\rho} + \nu\Delta u + f \qquad (2.5)$$

denoting by $\nu = \eta/\rho$ the *dynamic viscosity*. If the solution u does not explicitly depend on time, the profile is *stationary*. Otherwise, it is referred to as *transient*. A derivation of equation (2.5) from (2.4) under the assumption of a linear connection between shear and stress can be found in standard textbooks, e.g. Bruus, 2008; Landau and Lifshitz, 1991, or Batchelor, 1970.

In contrast to the advection-diffusion equation analyzed in chapter 1, (2.5) is nonlinear and consequently the solution space is not a linear space. Therefore, sums of solutions as well as their scalar multiples generally do *not* solve (2.5) anymore. In particular, this implies that a solution obtained for a specific geometry Ω cannot be mapped onto a scaled geometry. However, under certain conditions this is possible. Therefore, we write equation (2.5) in a dimensionless form by introducing a characteristic velocity scale U and a characteristic length scale L and denote all variables in respect to these coefficients. In detail, we define dimensionless variables

$$r' := \frac{r}{L} \qquad t' := \frac{tU}{L}$$

$$u'(r',t') := \frac{u(Lr',Lt'/U)}{U} \qquad p'(r',t') := \frac{p(Lr',Lt'/U)}{\rho U^2}.$$

Omitting external force densities and denoting $\nabla' = (\partial_{x'}, \partial_{y'}, \partial_{z'})$, equation (2.5) obtains the dimensionless form

$$\frac{\partial u'}{\partial t'} + (u'\nabla')u' = -\nabla' p' + \frac{1}{Re}\Delta' u' \qquad \text{with} \quad Re = \frac{UL\rho}{\eta} \qquad (2.6)$$

(2.6) only depends on one effective parameter: the dimensionless *Reynolds number Re*. The flow behaviour is strongly connected to this number. An example is given in Figure 2.4: A liquid flows through a meandering channel; the flow profile for different Reynolds numbers is shown. For low values (a), a stationary profile is obtained. Fluid elements initially situated near to each other should travel along parallel trajec-

2.1 Fundamentals

tories: a behaviour that is called *laminar* fluid flow. A characteristic change can be observed by raising the Reynolds number: the profile becomes transient and evolves in a chaotic manner as two arbitrary elements initially situated infinitesimally close to each other increase their distance exponentially. Such flow patterns are called *turbulent*.

The characterization of a flow profile can be carried out via a *streamline* analysis. A streamline at a time t is a curve γ_t which has the velocity vector $\boldsymbol{u}(t)$ of the profile as a tangent at each domain point. If a parameterization $\gamma_t(\lambda)$ is introduced for fixed time t, a streamline can be determined by the ordinary equation

$$\frac{d\gamma_t(\lambda)}{d\lambda} = \boldsymbol{u}(\gamma(\lambda),t) \quad (2.7)$$

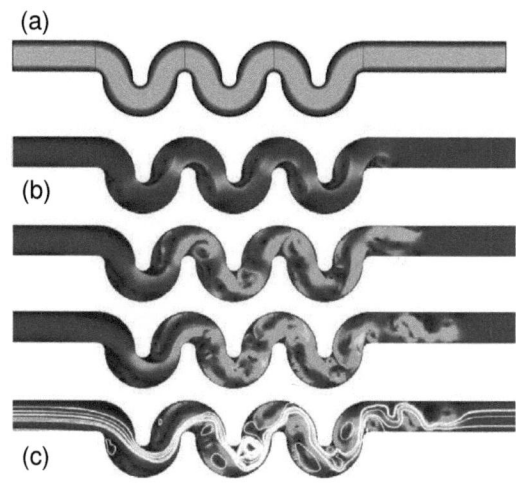

Streamlines coincide with the trajectories of liquid elements if a stationary flow is considered. An example is shown in Figure 2.4(c).

Figure 2.4: Flow through a meandering channel for different Reynolds numbers. (a) For low *Re* a stationary, laminar profile is obtained. (b) High *Re* instead leads to a turbulent, transient dynamic, the picture series shows the evolution of a chaotic flow. (c) Representation of a streamline plot for the high *Re*-regime.

2.1.4 The special case of microfluidics

It was already mentioned that the dimensionless Reynolds number $Re = UL\rho/\eta$ is an important parameter when classifying fluid flows. Fluid flows in microfluidic devices are usually on a velocity scale $U = \mathcal{O}(10^{-4}\,\text{m/s})$ while typical geometry details of the structure lie on a geometrical size scale $L = \mathcal{O}(10^{-4}\,\text{m})$. If we further consider a liquid with material parameters $\rho = \mathcal{O}(10^3\,\text{kg/m}^3)$ and $\eta = \mathcal{O}(10^{-3}\,\text{Pa s})$, it is

$$Re = \mathcal{O}(10^{-2}) \ll 1.$$

35

2. Microfluidic devices

This regime of very small Reynolds number is often referred to as *creeping flow regime*. Different terms in (2.5) may be estimated as follows

$$\frac{\mathcal{O}(\partial_t u)}{\mathcal{O}(\frac{\eta}{\rho}\Delta u)} = \frac{\rho \frac{U}{L/U}}{\eta \frac{U}{L^2}} = \frac{UL\rho}{\eta} = Re \ll 1 \tag{2.9a}$$

and

$$\frac{\mathcal{O}((u\nabla)u)}{\mathcal{O}(\frac{\eta}{\rho}\Delta u)} = \frac{\rho \frac{U^2}{L}}{\eta \frac{U}{L^2}} = \frac{UL\rho}{\eta} = Re \ll 1 \tag{2.9b}$$

According to (2.9a), inertia effects play only a minor role on the chosen scales; the time dependence of the solution can be omitted. Additionally, non-linear contributions vanish as well. We obtain the linearized *Stokes equation*

$$\eta \Delta u = -\nabla p. \tag{2.10}$$

Contrary to highly complex (non-linear) and involved full Navier-Stokes equation, one may hope to solve (2.10) together with the equation of continuity for highly symmetric problems. A famous example is the *Poiseuille flow* describing a fluid flow through a cylindrical tube of length ℓ and radius R imposing a pressure difference Δp between entrance and inlet and a 'no slip'-condition along the lateral area. Due to symmetry, the flow points along the symmetry axis and is given by the parabolic formula

$$u(r) = \frac{\Delta p}{4\eta\ell}(R^2 - r^2), \tag{2.11}$$

a schematic representation of the flow is shown in Figure 2.5. A similar result can be obtained for the flow between two parallel plates.

From a formal point of view, the discussion of microfluidic flow properties appears to be a lot easier due to the absence of non-linear dynamics and chaotic flows. On the other hand though, this leads to many practical problems since several phenomena ob-

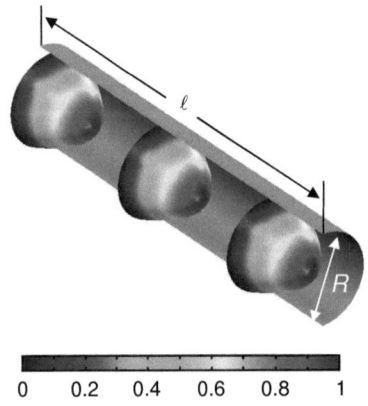

Figure 2.5: Poiseuille flow through a cylinder due to a pressure difference Δp at the ends of the pipe.

served on the macroscale cannot be employed on the microscale. A very important example is the mixing of e.g. a two-face flow.

2.1.5 Spherical objects dissolved in liquids

Small spherical objects dissolved in a liquid feel forces acting on them, arising either from the motion state of the liquid or from some external source. Gravity and buoyancy are typical examples of such external influences. For right now, all such effects shall be summarized in the external force F_{ext}. Additionally, a momentum transfer from the liquid to the dissolved object takes place. While external forces commonly act on the volume as a whole or some fraction of it, the fluid induced forces are transferred along the surface. If a constant fluid velocity is found along the surface, no force acts on the spherical objects; a force results from derivatives of the velocity components which is given be the stress tensor σ, with

$$\sigma_{ij} = \frac{\partial u_i}{\partial x_j} + \frac{\partial u_j}{\partial x_i}.$$

Additional forces originate from pressure gradients. The total force on a planar surface element of surface normal $\hat{n}$ is given by

$$f = (\sigma - p \cdot \mathrm{Id})\hat{n} \qquad (2.12)$$

with Id the identity matrix. The total force acting on an object dissolved in a fluid flow is given by integrating (2.12) along the interface between object and liquid. Since this includes the knowledge of the flow itself, analytic expressions are consequently rare. However, for the special case of a very small Reynolds number ($Re \ll 1$), the force a spherical particle of radius R moving at a velocity v feels in a homogeneous velocity field u is given by *Stokes drag law*

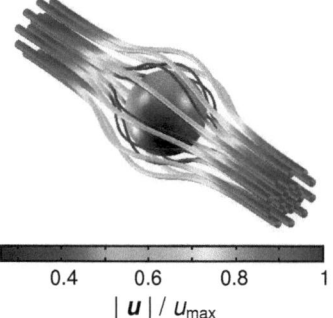

Figure 2.6: Streamline plot of a liquid flowing around a sphere.

2. Microfluidic devices

$$F_{stokes} = 6\pi\eta R(u-v), \quad (2.13)$$

a streamline plot is shown in Figure 2.6. If the particle diameter is much smaller than the geometrical size scale L, we can approach it by a point mass of density ρ_{part} and apply Newton's second law to calculate the particle behaviour. Due to the very small particle masses, inertia effects can be neglected; the particle velocity instantaneously equals the velocity of the liquid. In case of the presence of additional external forces, we may write (see Appendix A.2)

$$v = u + \frac{F_{ext}}{6\pi\eta R} \quad (2.14)$$

For higher Reynolds numbers this no longer holds. In this case, (2.14) can be replaced by the empirical *Khan-Richardson law* (Coulson and Richardson, 2003) which is valid along a wide range Reynolds numbers

$$F_{KR,i} = \pi R^2 \rho (u_i - v_i)^2 (1.84 Re_i^{-0.31} + 0.293 Re_i^{0.06})^{3.45} \quad (2.15)$$

with $\quad Re_i = 2R\rho \dfrac{|u_i - v_i|}{\eta}$

When dealing with a large number of particles, we do not investigate each individual component but introduce the particle concentration c as the number of particles per unit volume. Essentially, this approach follows the same guidelines that were already applied when going from the microscopic description of a liquid to the mesoscopic one. Due to the conservation of the total mass, a concentration change in respect to time can only originate from a space dependent mass flux j

$$\frac{\partial c}{\partial t} + \nabla j = 0. \quad (2.16)$$

Different kinds of mass fluxes can be considered. A velocity profile u e.g. induced by a moving liquid leads to a *convective flux* $j_{conv} = uc$. Left to themselves particles dissolved in a liquid do not remain resting but move due to thermal effects. An initially space dependent concentration will even out over time (compare Figure 2.6). This thermally activated mass transport can be described by a *diffusive flux*

2.1 Fundamentals

$j_{\text{diff}} = -D\nabla c$, with the diffusivity D. Superimposing the two transport mechanisms and employing equation (2.16) leads to the *advection-diffusion equation*

$$\frac{\partial c}{\partial t} - \nabla(D\nabla c) + \langle u, \nabla c \rangle = 0 \tag{2.17}$$

The diffusivity D may depend on space in different ways: on the one hand, it may change for very high concentrations, becoming a local function in the respect to the concentration $D = D(c(r))$ inducing an implicit space dependency. On the other hand, changes of the considered domain may significantly reduce the symmetry of the problem. A typical example can be found close to the channel wall of a microfluidic structure. Theoretical calculations predict a reduction of D to one third of the original value, an effect that can also be observed experimentally (Faucheux and Libchaber, 1994; Auge et al., 2009). For many practical applications though, it may be set at a constant value, depending on the properties of the dissolved objects and the carrier liquid. For small spheres of radius R, we set (Einstein, 1905)

$$D = \frac{k_B T}{6\pi\eta R} \tag{2.18}$$

with k_B the Boltzmann constant and T the absolute temperature.

Whereas the Reynolds number classifies the type of flow profile a similar dimensionless parameter may be introduced for advection-diffusion problems. The *Péclet number Pe* is given by the ratio between convection velocity and diffusion velocity. Denoting by L the diffusive length scale and by U velocity scale of the convection field, it is

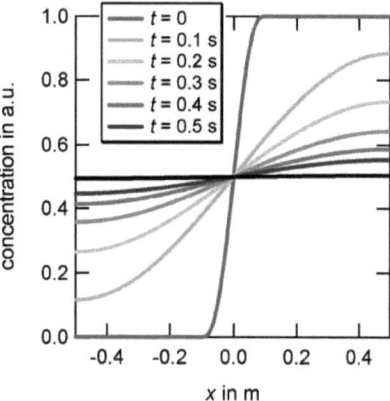

Figure 2.7: Evolution of the concentration profile on a one-dimensional domain for $D = 0.5$ m^2/s. The initial configuration (red line) begins to flatten out and reaches an equally distributed end state (black line).

$$Pe = \frac{UL}{D} \tag{2.19}$$

39

2. Microfluidic devices

Therefore, a high Péclet number corresponds to convection dominated systems, whereas a low value corresponds to diffusion dominated problems.

2.1.6 Boundary conditions

To completely specify a flow or diffusion problem, additional boundary conditions are needed. In order to proceed, it is necessary to define certain terminology and disambiguate common assumptions. We will consider a finite, open subdomain $\Omega \subset \mathbb{R}^n$ with boundary $\partial \Omega$ on which the Navier-Stokes equation (2.5) together with equation of continuity (2.3) is to be solved.

a) **'No slip'-condition**: A common assumption for liquids flowing close to a solid, non-moving wall that is due to the inner friction of the liquid: here, layers close to the wall move at the velocity of the wall itself, i.e. remain non-moving. A *'no slip'-condition* is therefore given by the homogeneous Dirichlet condition

$$u(r) = 0 \qquad \text{for } r \in \partial \Omega \qquad (2.20a)$$

b) **'Slip/Symmetry'-condition**: A 'no slip'-condition requires that the momentum of the liquid is completely transferred to the wall. If specially functionalized surfaces are introduced, a liquid motion may be observed, although a non-moving wall is considered due to small inter-atomic forces. In such cases the liquid can flow along the wall. Such a condition can be introduced in order to simplify the system, e.g. exploit certain symmetry conditions. 'Slip'- or 'Symmetry'-conditions introduce a flow parallel to the boundary. Denoting the surface normal vector by $\hat{n}$, such situations are modelled by the homogeneous Neumann condition

$$\langle \nabla u(r), \hat{n} \rangle = 0 \qquad \text{for } r \in \partial \Omega \qquad (2.20b)$$

c) **'Neutral'-condition:** The 'Neutral' boundary condition states that transport by shear stresses is zero across a boundary. This condition is denoted neutral since

2.1 Fundamentals

it does not put any constraints on the velocity and states that there are no interactions across the modelled boundary.

$$\eta\left(\nabla u(r)+(\nabla u(r))^T\right)\hat{n}=0 \quad \text{for } r \in \partial\Omega \tag{2.20c}$$

d) **'Inflow/Outflow'-condition:** Inlets and outlets of microfluidic are usually given by the Dirichlet conditions

$$u(r)=u_B(r) \quad \text{or} \quad p(r)=p_B(r) \quad \text{for } r \in \partial\Omega \tag{2.20d}$$

In the case of the advection-diffusion equation (2.17), it will be sufficient to define either the concentration (in case of inlets) or specify the mass flux along the boundary. For the latter, two choices are commonly used: Along the boundary, particles are (a) only transported by the convective velocity field u or (b) not at all, which are given by

a) **'Convective flow'-condition:** $\langle D\nabla c,\hat{n}\rangle=0 \quad \text{for } r \in \partial\Omega \tag{2.21a}$
b) **'Insulation'-condition:** $\langle -D\nabla c+uc,\hat{n}\rangle=0 \quad \text{for } r \in \partial\Omega \tag{2.21b}$

Specifying boundary conditions, a stationary profile is completely defined. In the transient case though, additional initial states $u(r,t_0)$ and $p(r,t_0)$ for an initial time t_0 must be declared.

2.1.7 Weak formulation

Following the example discussed in section 1.2, the weak formulations of the Navier-Stokes/Stokes and the advection-diffusion equation can be obtained. Here we denote the testfunctions for the variation of the velocity components by ψ_u and employ similar definitions for pressure p and concentration c. The choice of approximate function spaces is related to the maximum order of the weak derivatives found in the variational form. We denote subspaces by $V \subset H^1(\Omega)$ and $\tilde{V} \subset L^2(\Omega)$ for respective derivative order < 2 and $= 0$.

2. Microfluidic devices

Navier-Stokes equation:

$$\rho\left\langle\frac{\partial \boldsymbol{u}}{\partial t}, \psi_u\right\rangle_{L^2(\Omega)} + \rho\int_\Omega \langle \psi_u, (\boldsymbol{u}\nabla)\boldsymbol{u}\rangle\, d\boldsymbol{r}$$

$$= -\eta\int_\Omega \sum_j \left(\frac{\partial u_i}{\partial x_j} + \frac{\partial u_j}{\partial x_i} - p\delta_{ij}\right)\frac{\partial \psi_{u,i}}{\partial x_j}d\boldsymbol{r} - \int_{\partial\Omega}\psi_u\frac{\partial \boldsymbol{u}}{\partial \hat{n}}d\boldsymbol{r}$$

$$\int_\Omega \psi_p \cdot \nabla \boldsymbol{u}\, d\boldsymbol{r} = 0 \qquad\qquad \forall \psi_u \in V, \psi_p \in \tilde{V} \qquad (2.22\text{a})$$

Stokes equation:

$$\eta\int_\Omega \sum_j \left(\frac{\partial u_i}{\partial x_j} + \frac{\partial u_j}{\partial x_i} - p\delta_{ij}\right)\frac{\partial \psi_{u,i}}{\partial x_j}d\boldsymbol{r} = -\int_{\partial\Omega}\psi_u\frac{\partial \boldsymbol{u}}{\partial \hat{n}}d\boldsymbol{r}$$

$$\int_\Omega \psi_p \cdot \nabla \boldsymbol{u}\, d\boldsymbol{r} = 0 \qquad\qquad \forall \psi_u \in V, \psi_p \in \tilde{V} \qquad (2.22\text{b})$$

Advection-diffusion equation:

$$\left\langle\frac{\partial c}{\partial t}, \psi_c\right\rangle_{L^2(\Omega)} + \int_\Omega D\langle\nabla\psi_c, \nabla c\rangle\, d\boldsymbol{r} - \int_{\partial\Omega}\psi\langle\hat{n}, \nabla c\rangle\, d\boldsymbol{r}$$

$$+ \int_\Omega \psi_c\langle\boldsymbol{u}, \nabla c\rangle\, d\boldsymbol{r} = 0 \qquad \forall \psi_c \in V \qquad (2.23)$$

For mathematical reviews and analysis of (2.22) refer to Temam, 1984, Kreis and Lorenz, 1989, or Thomasset, 1981. A every thorough analysis of the advection-diffusion(-reaction) equation (2.23) can be found in the book of W. Hundsdorfer and J. Verwer, 2003.

2.1.8 Numerical stabilization and Petrov-Galerkin discretization

When dealing with advection-diffusion equations that are dominated by convective contributions (i.e. problems in the low Péclet regime), strong oscillations in the numerical solution can be found. An example is shown in Figure 2.8(a) or Figure 1.9 where fine details can be found in certain areas of the domain. Over time these features may propagate to other parts of the system and, therefore, the solution of the numerical scheme no longer resembles the solution of the original equation. Such local numerical instabilities may be overcome by increasing the mesh resolution. However, globally increasing the number of elements results in a large increase of numbers of freedom while numerical schemes introducing local mesh refinement (adaptive solvers) are still not well understood (Nochetto et al., 2009). Instead, in many cases a different type of stabilization, known as *artificial diffusion*, is employed, which may be understood as a generalization of the Galerkin discretization in section 1.2. Here per definition, the approximate solution u_h as well as approximate test functions ψ_h are chosen from the same finite-dimensional space V_h. In contrast, the *Petrov-Galerkin method* still employs Lagrangian elements for the construction of the approximate solution, although for the testfunctions functions that are signed and weighted in respect to the convective flow are used.

In principle, such methods increase the diffusion within the system, where the stabilizing parameters depend on the local mesh dimension h and the norm of the convective flow $|\boldsymbol{u}|$. In the simplest case, the *isotropic artificial diffusion*, an additional term

$$D_{art} = \frac{1}{2} h |\boldsymbol{u}|$$

is added to the diffusion constant D. Though in most cases oscillations are damped and instabilities vanish, the solution might not be satisfactory since it no longer solves the original equation. The

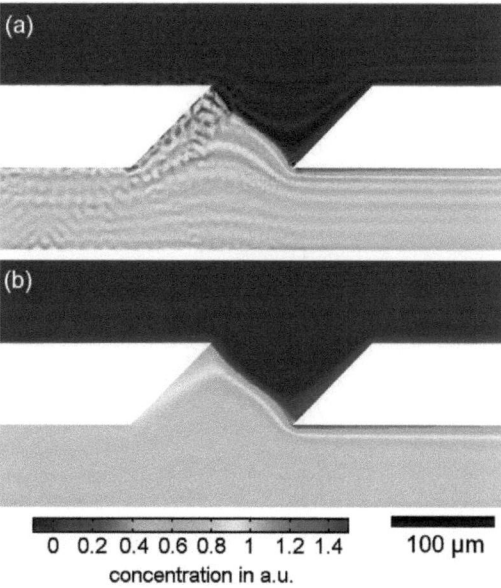

Figure 2.8: Numerical instabilities in convection dominated systems. A concentration travels in the lower part of the geometry, (a) shows a solution obtained by a Galerkin approach. Instabilities manifest in oscillations that may also lead to large negative concentrations (~ − 0.1). (b) presents a solution calculated in a Petrov-Galerkin framework. No more wiggles can be found.

2. Microfluidic devices

Petrov-Galerkin method does not change the original equation, but modifies the space of testfunctions. Instead of using the hat functions Λ_i, an additional term is added. We write

$$\tilde{\Lambda}_i = \Lambda_i + \kappa \sigma_i \qquad (2.24)$$

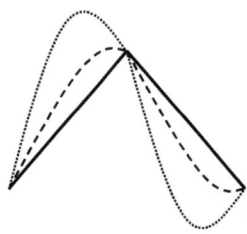

Figure 2.9: Basis functions with different values for σ_i in comparison to a hat function.

with a direction dependent parameter κ and an addition weight σ_i. Examples of such functions are shown in Figure 2.9. For the advection-diffusion equation, a typical choice is to set

$$\tilde{\psi} = \psi + \delta'(\delta' > 0)\langle \mathbf{u}, \nabla \psi \rangle$$
$$\text{with} \quad \delta' = \frac{\delta h}{|\mathbf{u}|} - \frac{c}{|\mathbf{u}|^2}$$

denoting by δ a tuning parameter. This method is also known as *streamline diffusion* or *streamline upwind Petrov-Galerkin* (SUPG). An additional diffusion is introduced via the testfunctions, corresponding to an artificial diffusion tensor of the form

$$D_{sd,ij} = \delta'(\delta' > 0) u_i u_j$$

Changing the testfunction space results in numerical stabilization as is apparent from Figure 2.9(b). For further information on these methods, see e.g. Neslitürk and Harari, 2003 or Franca et al., 1992.

2.2 Particle separation by a hydrodynamic switch

In order to understand some features of the behaviour of small magnetic beads in microfluidic devices, we will investigate the separation properties of the microfluidic geometry shown on the left side of Figure 2.10(a). Two microfluidic channels of a width of 80 μm run parallel to each other with a distance h and are connected via an additional channel segment cutting both channels under an angle α. For the inflow velocity in both entrances, a parabolic Poiseuille profile according to equation (2.11) is assumed where different maximum values at upper and lower entrance are chosen. Denoting these inflow velocities by u_{up} and u_{down} at the inlets A and B, respectively, a relation factor ξ can be defined by

$$\xi = \frac{u_{up}}{u_{down}}. \tag{2.25}$$

In particular, an equilibrium value ξ_0 may be identified where no volume flux from the lower part B of the geometry to the upper part A can be found. The influence of ξ on the flow behaviour is shown in Figure 2.10(b). The proposed geometry acts as a hydrodynamic switch: the flow direction can be adjusted by specific choices of ξ. In

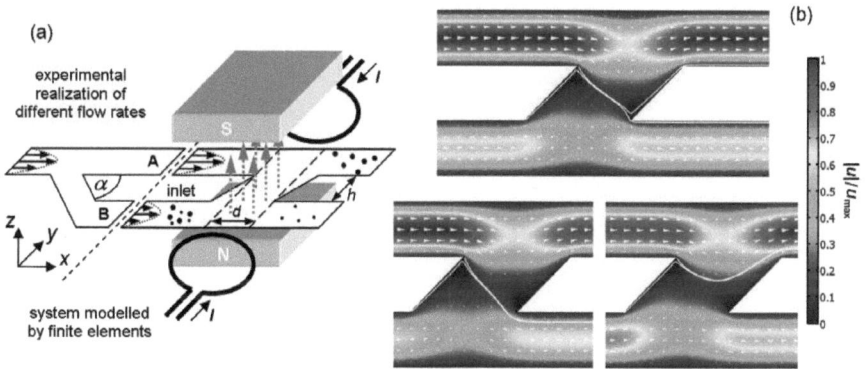

Figure 2.10: Investigated microfluidic separation device. (a) Schematic representation of the structure and the external field contributions: the geometry consists of two flow inlets of different maximum velocities and two outlets. The magnetic field can be divided into an inhomogeneous contribution generated by two ring-shaped wires and a homogeneous z-directed field. (b) Hydrodynamic properties of the device, white lines indicate the streamline behaviour in the junction area. For a certain relation factor ξ no volume exchange between the channel segments can be found (top). Situations for $\xi > \xi_0$ and $\xi < \xi_0$ are shown in the bottom left and bottom right inset, respectively.

45

2. Microfluidic devices

detail, we obtain a flow from B to A or A to B if either $\xi < \xi_0$ or $\xi > \xi_0$ is chosen, respectively. Additionally, the switch is shut for $\xi = \xi_0$.

According to (2.14), a magnetic particle passing the separation area follows the streamline profile of the fluid flow if no further external forces act on the particle. To manipulate magnetic beads along the junction area, an inhomogeneous magnetic field is required, which can be introduced by current leading wire geometries on the microscale. Many designs have been proposed (see e.g. Lee et al., 2001), in our approach, we consider a wire geometry of two opposite rings as indicated. The design of the wire configuration follows a guideline according to Weddemann, 2006, Figure 2.11 shows a finite element simulation of the behaviour of the magnetic field between the two rings with a current of 1 mA in each component.

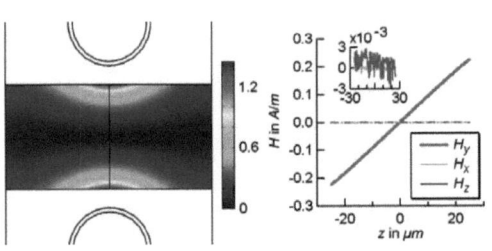

Figure 2.11: Inhomogeneous magnetic field created by ring-shaped electric wires with a current of 1 mA. (a) shows the norm of the field, (b) different components of *H* in the centre of the geometry. Apart from the component H_y all other contributions are at the scale of numeric noise. Therefore, a force only results from the derivative $\partial_z H_y$.

If only electric currents are taken into account, we can easily conclude that a particle will always feel an attractive force pointing towards the nearest wire: as the magnetic moment m_{part} of the particle aligns with the inhomogeneous magnetic field, it is $m_{part} \parallel H_{ext}$. Therefore, we may write

$$m_{part} = |m_{part}| \frac{H_{ext}}{|H_{ext}|}$$

The force acting on a magnetic particle is given by (compare section 4.2)

$$F_{part} = \mu_0 (m_{part} \nabla) H_{ext}. \qquad (2.26)$$

As $\varepsilon_{ijk} \partial_j H_{ext,k} = 0$, it is

$$H_{ext,j} \partial_j H_{ext,i} = H_{ext,j} \partial_i H_{ext,j} = \frac{1}{2} \partial_i H_{ext,j}^2$$

$$\Rightarrow F_{part} = \mu_0 (m_{part} \nabla) H_{ext} = \mu_0 \frac{|m_{part}|}{|H_{ext}|} \nabla H_{ext}^2$$

2.2 Particle separation

As per definition, the gradient vector ∇H_{ext}^2 points into the direction of the highest increase, the direction of the current density itself, which is due to the radially decaying field $\sim r^{-1}$. Therefore, to achieve forces pushing particles from one side of the geometry to the opposite one, it is necessary to uncouple the direction of the magnetic moment from the inhomogeneous wire field. For this purpose, the whole setup is brought into a strong homogeneous magnetic field pointing perpendicular to the plain of the microfluidic structure, compare Figure 2.10(a). Such a field does not exert any force onto the particles but fixes the magnetization direction along the z-axis and we may write $m_{part} = m_z \hat{m}$. Thus, formula (2.26) describing the force onto the particle simplifies to

$$F_{part} = \mu_0 (m_{part} \nabla) H_{ext} = \mu_0 m_z \frac{\partial H_{ext}}{\partial z} \quad (2.27)$$

The force in y-direction is determined by the derivative $\partial_z H_y$. The numerical simulations presented in Figure 2.11 show that this component is the only derivative contributing in the centre of the channel geometry; all other components are on the size scale of numerical noise. Combing equation (2.27) with equation (2.14), the velocity of particles passing the separation area depends on two particle properties: the size and the magnetization. Therefore, the microfluidic geometry can be used as a separation device for particles: We assume that all particles enter the geometry through the lower entrance B. For the considerations here, the particle concentrations are chosen sufficiently low so that particle-particle interactions can be omitted. As a carrier liquid, we choose water at room temperature, which provides the parameters T = 300 K, ρ = 998.2 kg/m³ and η = 1.002 mPa s. If we denote the inlet boundaries by $\partial\Omega_A$ and $\partial\Omega_B$ and summarize the exits by $\partial\Omega_{ex}$, the complete set of equations is

Concentration dynamics:

$$\frac{\partial c}{\partial t} - \frac{k_B T}{6\pi\eta R} \Delta c + \langle v, \nabla c \rangle = 0 \quad \text{with } v = u + \frac{F_{mag}}{6\pi\eta R} \quad \text{on } \Omega$$

$c = 1 / c = 0$	inlet concentration	on $\partial\Omega_A / \partial\Omega_B$
$\langle \hat{n}, D\nabla c \rangle = 0$	only convective outflow	on $\partial\Omega_{ex}$
$\langle \hat{n}, -D\nabla c + vc \rangle = 0$	insulation	elsewhere

2. Microfluidic devices

> where the velocity u follows from
>
> $\Delta u = \eta \nabla p$ and $\nabla u = 0$ on Ω
>
> $u_x = 4(1-s)s \cdot u_{max}, u_y = 0$ Poiseuille profile with   on $\partial\Omega_A$
> $u_x = 4(1-s)s \cdot \xi u_{max}, u_y = 0$ parametrization $s \in [0,1]$ on $\partial\Omega_B$
> $\eta\left(\nabla u(r) + (\nabla u(r))^T\right)\hat{n} = 0$ Neutral-flow-condition   on $\partial\Omega_{ex}$
> $u = 0$ 'No slip'-condition   elsewhere

Since the flow profile is stationary, we focus on stationary solutions of the equation system only. For small particles, all equations are discretized using quadratic Lagrangian elements except for the pressure p which is approximated via linear functions. Analyzing particles of a size exceeding ~ 0.5 μm leads to convection dominated systems; diffusive fluxes play a minor role. Therefore, for these particle sizes a Petrov-Galerkin approach is employed for the numerical discretization of the Stokes equation to guarantee convergence.

The results of the simulations are shown in Figure 2.12, which presents the relation between particles leaving the geometry through the upper exit and the inflow concentration with respect to the particle size is shown. If all particles leave through the lower exit a value of "1" results, whereas a value of "0" is obtained if all particles leave through the lower exit instead. Therefore, these simulations determine what particles of which sizes can be separated for the given set of parameters.

The bigger the junction area ($d \to 320$ μm, $h \to 320$ μm), the more important the diffusive effects acting on the particles become. In these cases, very small particles can always be found at both exits if a relation factor of $\xi = 1.5$ is chosen, resulting in a bad separation yield of the device. However, an increase of the velocity inflow ratio ξ can always suppress particle diffusivity (Figure 2.12(b), (d), (e)). Thus, the proposed microchannel geometry is suitable for even particles on the size scale of several 10s of nanometers. Furthermore, in all cases the size interval of particles that can be found in both exits is very narrow for high ξ. Therefore, it is also possible to separate particles with desired accuracy by a proper adjustment of the fluidic and geometric parameters.

2.2 Particle separation

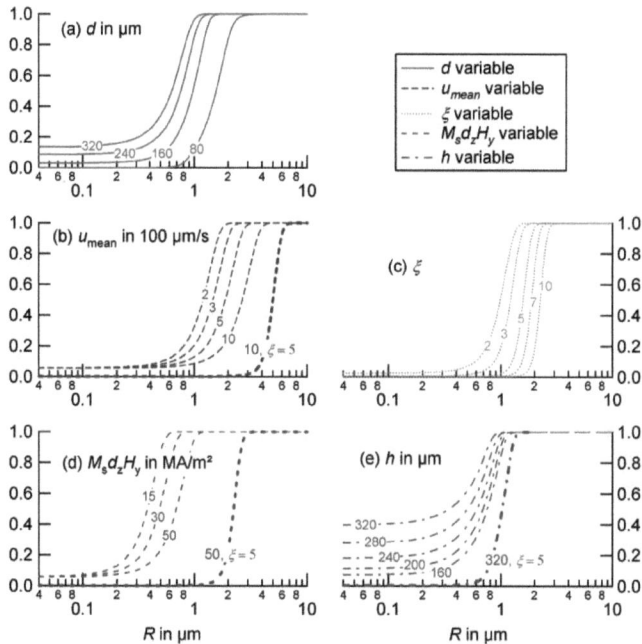

Figure 2.12: Relation between mass flow through upper and lower exit. A value of "1" corresponds to all particles reaching the upper part of the geometry, a value of "0" indicates no particle switching from the down to up. The plots show the dependence of the separation properties on different system parameters. Values not explicitly given coincide with the reference values $d = 200$ μm, $h = 80$ μm, $\xi = 1.5$, $u_{down} = 100$ μm/s and $M_s\partial_z H_y = 10000$ kA/m².

The experimental verification of the separation device has been realized by F. Wittbracht in the framework of his master thesis (Wittbracht, 2009). All experiments were performed with an optical microscope and an attached CCD-camera. The setup enables the simultaneous recording of images and current through the conducting lines at 8 frames per second. For the creation of a homogeneous external magnetic field in z-direction, a cylindrical coil was used. The magnetic gradient field is generated by conducting lines on the microfluidic chip as shown in Figure 2.10. Samples are prepared using optical laser lithography and magnetron sputtering for gold conducting lines and UV-lithography for microfluidic channels. The epoxy-based resist SU-8 25 is used as channel material due to its mechanical and chemical stability (Lin et al., 2002). As the fluid flow is generated by hydrostatic pressure, a fixed velocity

2. Microfluidic devices

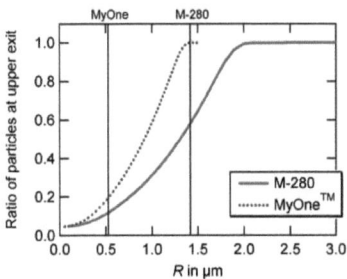

Figure 2.13: Theoretical prediction of the separation properties for the experimentally employed M-280 and MyOne™.

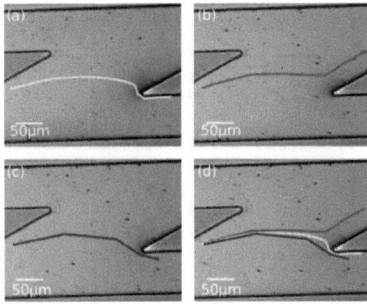

Figure 2.14: Typical bead trajectories through the separation region for different bead species and currents. (a) M-280 beads enter and leave the separation region through the lower channel at 0 mA. (b) These beads can be dragged to the upper channel by applying a current of 180 mA to the conduction lines, while MyOne™ enter and leave through the lower channel. (d) shows a superposition to ease the com-

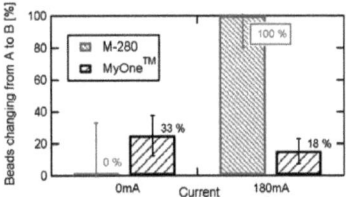

Figure 2.15: Comparison between bead flows from B to A for different electric currents and bead species.

ratio ξ is difficult to realize experimentally when using two inlet reservoirs. Therefore, in experiments the velocity ratio is implemented using one inlet reservoir and a channel geometry that splits into two channel branches as shown schematically in Figure 2.10, with an angle $\alpha = 26{,}6°$ obtained also from finite element calculations. For size separation experiments, a mixture of Dynabeads® MyOne™ (1.05 μm diameter) and M-280 (2.8 μm diameter) superparamagnetic beads is used. Both bead species have narrow very size distributions (CV $\approx$ 1.9 %) and comparable susceptibilities (Fonnum et al., 2005). The homogeneous magnetic field is adjusted to 557 Oe which is the maximum of the described setup. For these field values, both bead types are almost completely saturated. This verifies the assumption of the simulations of saturated magnetic beads and therefore validates (2.27). The current applied through the conducting lines is varied during experiments. Employing the particle parameters for the simulations similar plots can be obtained as shown in Figure 2.13.

The maximum velocity value u_{max} is 400 μm/s. In this case, the geometry parameters are chosen as $d = 160$ μm and $h = 80$ μm. For a separation at higher flow rates, the geometry parameters need to be adjusted in respect to Fig. 2.12. Therefore, a separation is theoretically possible at any given flow rate and only limited by experimental constraints. Due to the experimental realization, the bead solution fills both channels and the analysis

2.3 Particle transport

of experimental results needs to be based on single bead tracking. Manual tracking is achieved using ImageJ, 2009, and the MTrackJ plugin, 2009. In order to grant comparability of the data, tracks of beads at similar positions in the channel are evaluated (Figure 2.14(d)). M-280 and MyOne™ beads enter and leave the separation region through the lower channel (Figure 2.14(a)) if no current is applied to the (ring-shaped) conducting lines. When a current of 180 mA is applied, it leads to a magnetic force which drags the M-280 beads to the upper channel exit (compare Figure 2.14(b)). At the same current, MyOne™ beads enter and leave through the lower channel exit as shown in Figure 2.14(c). Figure 2.15 is obtained by observing the behaviour of bypassing beads and shows the ratio of beads flowing from B to A. It is clearly demonstrated that at a current of 180 mA the M-280 can be completely separated from the MyOne™. The small degree of impurity (18 %) can be attributed to a factor ξ too small to completely suppress a volume flow from B to A.

2.3 Transport properties for the low Péclet-regime

One predominant issue for μTAS devices is the precise magnetic transport of the beads to functional device sections such as the reaction chamber or the separation site of the microfluidic chip. The transport of magnetic markers in a microfluidic channel is influenced by hydrodynamic and magnetic forces. Brownian motion is also important for nanoparticles. While a high diffusion is desirable for mixing in a reaction chamber, this is not the case for continuous flow separation or directed transport to other parts of the fluidic device. A narrow spatial particle distribution stable against diffusion is preferable for the latter (Pamme et al, 2003). This can be achieved by different means: one possibility is hydrodynamic guidance as has been proposed in various works, where a focusing of the particle distribution is achieved by additional flow inlets (Jahn et al. 2004; Dittrich and Schwille, 2003) or by enforcing high Reynolds numbers on the microscale (Carlo et al., 2007). Other methods exploit e.g. acoustic (Shi et al., 2008) or dielectrophoretic effects (Ravula et al., 2008). However, the examples given above need either a higher number of inlets or additional physical phenomena, such as non-linear flow behaviour, acoustics or electromagnetics, all of which increase the complexity of the device. Further, for many devices the applicability to the small particle limit is often not discussed, though becomes increasingly important, especially if the focused particle beam must be transported to other sites of a lab-on-a-chip system.

2. Microfluidic devices

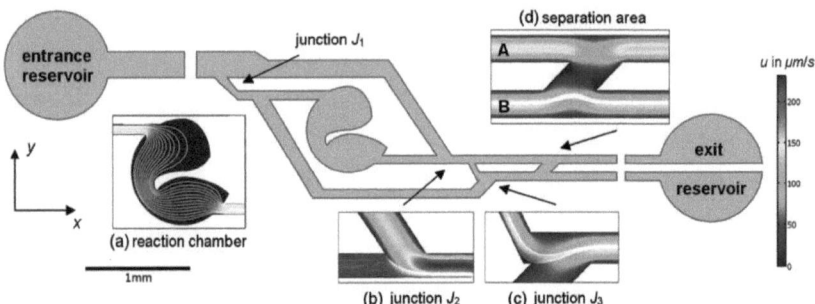

Figure 2.16: Microfluidic lab-on-a-chip geometry consisting of one inlet reservoir (left) and two outlets (right). In the centre, a reaction chamber for e.g. particle supply or chemical reactions is embedded. The streamline course of the product leaving the reaction chamber can be found in the subplots and determines the particle trajectory for particles of radius down to several 100 nm. The two junctions J_1 and J_3 can be used to manipulate the flow in the separation area on the right side in the channel segments A and B (see below). At junction J_2, the focusing of the particle beam takes place. Additionally, J_2 is optimized t to suppress particle diffusion, preventing particle flow into the channel segment A.

The scope of this section is to integrate the separation device introduced in the preliminary section into a microfluidic structure which also fulfills the following requirements:

(1) consists of only one inlet and creates the velocity ratio in the separation site by microfluidic flow guidance
(2) enables chemical reactions along a reaction site, where diffusive transport is dominant
(3) grants a stable transport from the reaction chamber to the separation device

Here we will try to achieve a very narrow particle beam which in turn enhances the separation yield. The analyzed microfluidic lab-on-a-chip geometry is shown in Figure 2.16. As specified, it consists of only one entrance which splits into three channel branches. The middle branch passes through a reaction chamber; its shape is optimized in order to achieve a high percolation (Figure 2.16(a)). Additionally, geometry adjustments prolong the average duration of incoming species in the chamber so that sufficient reaction times can be granted for applications. The separation structure can be found at the right end of the device.

It was already demonstrated that particles can be separated from each other with respect to different properties. For a given property (e.g. size), the parameter ξ determines which values of this property can still be separated from each other. Therefore, the degree of adjustment of ξ is important for the adaptability of the system to differ-

2.3 Particle transport

ent particle species and will be addressed below. Apart from this parameter, a narrow spatial particle distribution flowing from the reaction chamber to the separation area is strongly recommended and will lower the size difference of particle species that can still be separated from each other. Increasing separation yield or analysis resolution by proper spatial preconditioning has been reported in previous works (Pamme et al., 2003). The proposed geometry therefore also fulfills the purpose of focusing particles into a narrow particle beam by iterative separation and recombining the fluid flow via several channel junctions (J_1, J_2 and J_3, Fig. 1). The main influence limiting the functionality of the device is particle diffusion which decreases the stability of transport and separation properties. Therefore, the main problem is finding a lower bound for the particle size.

The simulations are carried out in exactly the same manner as was done for the analysis of the separation device. Again, velocity components and particle concentration are approximated by quadratic Lagrangian elements, the pressure is discretized by linear functions. For sufficiently small particles, calculations are carried out in a Galerkin-framework, whereas particles of sizes > 500 nm are treated by a Petrov-Galerkin approach. The insets of Figure 2.16 show simulation details of the hydrodynamic properties at important places along the device. It should be immediately pointed out that the fluid velocity in the reaction chamber is strongly reduced in comparison to the maximum inflow velocity. Therefore, the duration time of particles in this area is enhanced, thereby increasing the time scale on which chemical reactions etc. can occur. Averaging the fluid velocity and exploiting the linearity of the Stokes equation, the average time $\langle t \rangle$ a particle spends in the chamber may be estimated by

$$\langle t \rangle = g_{\text{lat}} u_{\text{max}}^{-1}, \tag{2.28}$$

with a geometry dependent parameter given by $g_{\text{lat}} = 0.09$ m for the geometry shown in Figure 2.16. When considering the motion of big particles ($r > 200$ nm), diffusive effects play a minor role. Since in these cases no additional external forces act on the particles, the dynamic behaviours result from the streamline plot of the fluid flow according to equation (2.14). The results can be found in Figure 2.16. Particles leaving the reaction chamber are focused into a narrow particle beam (Figure 2.16 (b)) and follow the flow profile to the lower section of the separation area. Here the particle beam is pushed upwards, nearer to the separation junction (Figure 2.16(c), (d)).

2. Microfluidic devices

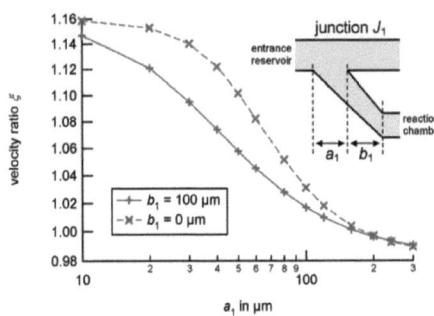

Figure 2.17: Influence on the velocity ratio on ξ of the junction J_1 for different junction geometries. A change of J_1 enables adjusting of ξ without influencing the position of the particle beam.

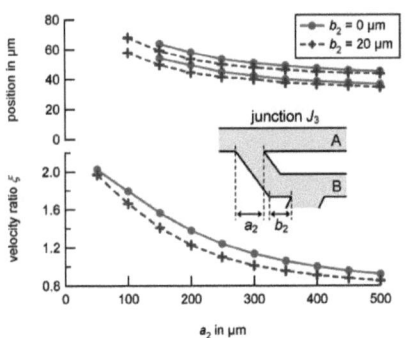

Figure 2.18: Influence on the velocity ration ξ of the junction J_3 for different junction geometries. A change of J_3 enables adjusting of ξ together with the position of the particle beam.

Velocity ratio ξ and spatial distribution can directly be controlled by adjusting the two small junctions J_1 and J_3. Therefore, a change of J_1 affects the ratio ξ as shown in Figure 2.17, where ξ is plotted in respect to the inlet size for two different junction shapes. The spatial distribution remains almost constant. Both parameters may be influenced by changing the third junction J_3. The upper subplot of Figure 2.18 shows the influence of the inlet size on the spatial distribution. The two lines here indicate the highest / lowest streamline starting in the reaction chamber and therefore the spatial bounds of the particle beam. We find that the width of the distribution remains almost constant but the position in the channel (width 80 µm) changes. An increase of ξ for a decreasing distance between separation junction and beam can be reported. Applications with small particles strongly benefit from this correlation, as diffusion can be suppressed by a higher value of ξ. Small geometry changes are therefore suitable for positioning the particle beam in the flow profile and also adjusting the velocity ratio. According to the results of section 2.2, Figure 2.10, a change of ξ from 1 to 5 can strongly change the separation yield at the separation site. Therefore, a change of the junction J_3 is suitable to achieve a value of ξ and therefore to adjust the flow behaviour to specific particles. Junction J_1 can be used for a precise tuning of the fluid flow.

For large particles, the second junction area J_2 ensures the focusing of the particle beam. If smaller particles are used, the fluid flow through J_2 also suppresses particle diffusion to the channel segment A of the separation device (Figure 2.19(a), (b)).

2.3 Particle transport

However, a narrow spatial particle distribution in channel segment B can only be found for particles of radius down to $R = 25$ nm. For smaller particle size, the distribution flattens out in the separation junction (Figure 2.19(a), (b)), lowering the resolution of the separation. Further stabilization requires higher inflow velocities (Figure 2.19(c), $u_{max} = 1$ mm/s). To analyze adaptability to particles on the nanoscale, we calculate the *normalized local Péclet number*

$$Pe_{loc} = \frac{|u|}{u_{max}} \left(\frac{\ell}{RD}\right)^{-1}, \qquad (2.29)$$

with the diffusive length scale ℓ. High Pe_{loc} indicates a convection-dominated regime while a low value corresponds to a diffusion-dominated one. Choosing $\ell = 40\,\mu m$ as half the channel width and restricting the evaluation of Pe_{loc} to the area of the desired beam position, Figure 2.19(d)) is obtained. For a narrow distribution, a high Péclet number is needed along channel segment B, which can be exploited to find a lower bound for the maximum velocity. Demanding a threshold of

$$Pe = \|Pe_{loc}\| R u_{max} > 1$$

the following estimation holds

$$u_{max} > \frac{1}{R\|Pe_{loc}\|} \qquad (2.30)$$

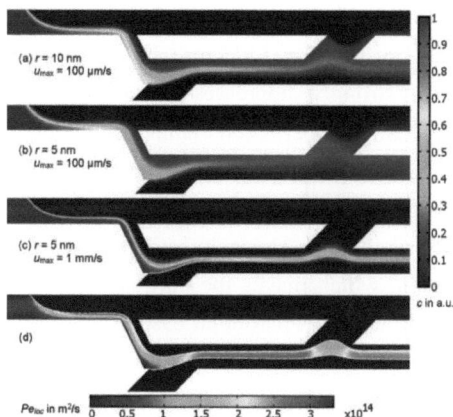

Figure 2.19: Influence of the particle diffusion on the particle beam behaviour for (a) $R = 10$ nm, $u_{max} = 100\,\mu m/s$, (b) $R = 5$ nm, $u_{max} = 100\,\mu m/s$, (c) $R = 5$ nm, $u_{max} = 1$ mm/s. (d) shows the plot of Pe_{loc} along the beam line of (c); high values correspong to convection-, low to diffusion-dominated regimes.

Equation (2.30) together with equation (2.28) can be used to calculate the working range of the device, i.e. the maximum duration time $\langle t \rangle_{max}$ of the particles in the reaction chamber in respect to the particle size as

$$\langle t \rangle_{max} = \frac{g_{lat}}{v_{rel}} \frac{\ell}{D} = \frac{g_{lat}}{v_{rel}} \frac{6\pi\eta\ell}{k_B T} R \qquad (2.31)$$

2. Microfluidic devices

where we set $\|\boldsymbol{u}\|/u_{\max} = v_{\mathrm{rel}}$ and v_{rel}, again a geometric parameter. For the geometry shown in Figure 2.16, we obtain from simulations $v_{\mathrm{rel}} = 1.5$. Therefore, the maximum particle duration in the reaction chamber is proportional to the particle radius.

For the experimental comparison, the microfluidic structure was realized by B. Eickenberg and F. Wittbracht. Dynabeads MyOne™ are used to visualize the flow behaviour. It needs to be pointed out that the magnetic properties are of no importance here. The creation of the microfluidic structure coincides with the procedure already explained in section 2.2 (a picture is shown in Figure 2.20(a)). The fluid flow is generated via hydrostatic pressure by filling the entrance reservoir with an aqueous particle solution with a concentration of 0.02 mg/ml.

To acquire flow information along the whole structure by particle imaging velocimetry, the homogenous solution is filled into the geometry entrance instead of the reaction chamber. The resulting streamlines within junctions J_2 and J_3 are shown in Figure 2.20(b). White lines correspond to trajectories of particles passing the reaction chamber, others are coloured in black. We find the theoretically predicted focusing effect of junction J_2 (Figure 2.20(b)), the volume flux coming from the entrance reservoir strongly suppresses the outlet of the reaction chamber and no particles from the reaction chamber reach channel segment A. At junction J_3 the focused particle beam is lifted toward the centre of the channel B. The general behaviour coincides with the theoretical prediction, however, in detail the streamline profile differs slightly from the calculation. These deviations can be attributed to a very low velocity along this region, which makes the particle behaviour very sensitive to small perturbations in the flow. In this particular case such perturbations can be provoked by a slight undercut profile of SU-8 microchannel as reported in (Liu et al., 2005). The particle trajectories along the separation site (Figure 2.20(c)) match the numerical simulation very well.

In order to experimentally validate the functionality of the device in the low Péclet number regime, the flow velocity of the particles is analyzed. The hydrodynamic pressure generates an inflow profile with $u_{\max} = (435 \pm 30)\ \mu\mathrm{m/s}$. An average velocity of $u = 20\ \mu\mathrm{m/s}$ is measured along the reaction chamber. The geometrical constant g_{lat} can be evaluated as $g_{\mathrm{lat}} = (0.1085 \pm 0.0075)\ \mathrm{m}^2$ and is therefore in strong agreement with the theoretical model. The small deviation results from reasons similar to those already discussed above. This is an experimental verification of the validation of equation (2.28), together with the streamline behaviour the estimation (2.31) has therefore been proven from experiments.

2.3 Particle transport

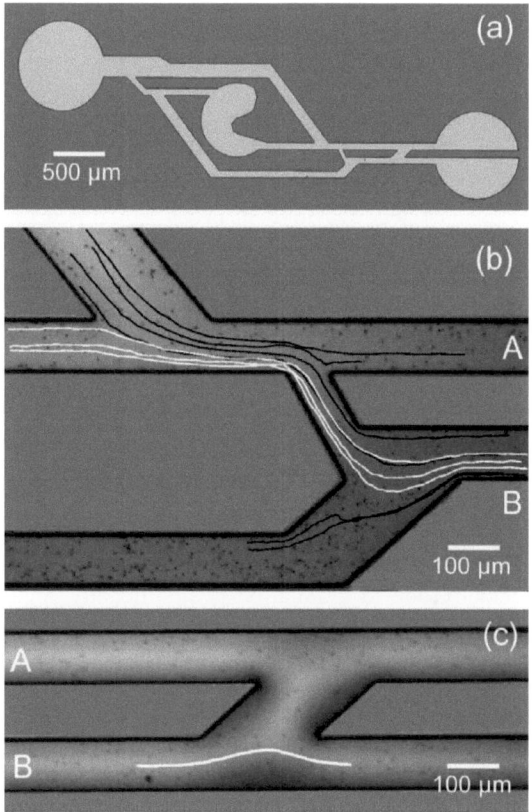

Figure 2.20: Microscope images of the experimental realization. (a) shows an overview of the whole structure, (b) typical streamline plots in the junctions J_2 and J_3. White lines pass the reaction chamber, other trajectories are coloured in black. (a) shows the streamline plot in the separation site.

2. Microfluidic devices

2.4 Microfluidic gravitational positioning system

At this stage, particles have undergone most of the lab-on-a-chip procedures and, following the microfluidic outline given in the introduction, they now reach the detection area. We will still see in chapter 5 that a measurable signal can only be found for a sensor-particle distance of below some micrometers due to a rapid decay of the particle stray field. Therefore, the remaining task of the µTAS-device is to position the particle close to the sensor i.e. on the bottom of the channel. As was already discussed in section 2.2, one method for doing this is to employ additional electric wires located close to the magnetically sensitive surfaces, which pull particles in their direction. Such strategies have been thoroughly investigated (Mikkelsen and Bruus, 2005, or Krishnau et al., 2008). However, they have several disadvantages: due to the two-dimensional current distribution, magnetic particles must be very near to the target region, i.e. the actual sensor, in order to be attracted. Therefore, the capture rate is comparatively small. On the other hand, additional electric components increase the complexity of the device and also lead to strong particle-particle interactions which entail agglomerations (Sinha et al., 2009). Therefore, we try to pursue a different strategy here.

In many microfluidic systems, it is not necessary to study gravitational effects, since the particle velocity due to gravitation is significantly smaller than the fluid flow itself. However, denoting by ℓ_z the length scale in z-direction of the geometry, by g the gravitational constant and by U the (convective) velocity scale, this is no

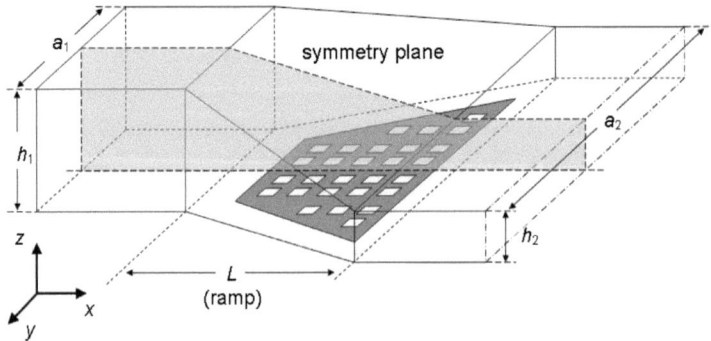

Figure 2.21: Schematic representation of the geometry. A rectangular microfluidic channel of height h_1 and width a_1 changes over a length L into a rectangular channel of height h_2 and width a_2. The particle target region e.g. a coated sensor array is place on the bottom of the channel section of decreasing height (ramp)

2.4 Particle positioning

longer true, if the Froude number $Fr = (g\ell_z/U)^{-1/2}$ is on the size scale of $\mathcal{O}(1)$. In this case the resulting particle motion can be strongly affected by gravity and buoyancy effects, and therefore may no longer be ignored. The microfluidic geometry investigated for this purpose is schematically shown in Figure 2.21: a rectangular shaped channel of height h_1 and width a_1 changes over a length L into a channel of height h_2 and width a_2. The channel segment of decreasing height will be addressed as *ramp* in the following. The target area can be found along the bottom of the ramp section,. In the actual application, this is an array of magnetoresistive sensors with a bio-functionalized surface enabling specific binding between surface and magnetic markers if the correct linker molecule is present in the sample. For later purposes, we define the cross section ratio χ by the relation of the area of the cross sections of the ramp exit and the ramp entrance

$$\chi = \frac{A_2}{A_1} = \frac{h_2 a_2}{h_1 a_1}. \tag{2.26}$$

For the numerical calculations, we consider particles of a diameter $d = 1$ μm and a density $\rho_{part} = 2500$ kg/m³ coinciding with the particles used in experimental realization. Gravitation enters the advection-diffusion equation via an additional velocity term

$$\frac{\partial c}{\partial t} + D\Delta c - vc = 0 \quad \text{with} \quad v = u + \frac{2R^2}{9\eta}(\rho - \rho_{part})g\hat{z} \tag{2.27}$$

with u a solution of the Stokes equation for density ρ and viscosity η, D the diffusion constant and g the gravitational constant. Due to the symmetry of the system, only one half is modeled (indicated by the grey plane, Figure 2.21), imposing a 'Symmetry'-condition $\langle \hat{n}, \nabla u \rangle = 0$ along plane of symmetry. Additionally, we assume convective flow condition $\langle \hat{n}, D\nabla c \rangle = 0$ along the target region, corresponding to the assumption that a particle reaching the ground will immediately bind to the biocoating. The resulting set of equations is solved again in Galerkin / Petrov-Galerkin framework. Due to calculations constraints, the size of the device is limited within simulations. This is due to the necessity for a high mesh resolution of the finite element discretization at the bottom of the channel which leads to a large number of degrees of freedom for structures of a length $L > 1.5$ mm i.e. high aspect ratios of the employed

2. Microfluidic devices

geometry. The resulting concentration profiles on the bottom of the ramp for the geometry parameters $h_1 = 50$ μm, $a_1 = 80$ μm, $h_2 = 20$ μm, $a_2 = 200$ μm, an inflow velocity $u_{in} = 200$ μm/s and different values of $L = 400, 800, 1200$ μm are shown in Figure 2.22 (upper subplot of each graphic). For sufficiently high L, a local minimum and maximum in the distribution can be found. Therefore, this yields constraints on the sensor position on the bottom of the microfluidic device: If a homogeneous surface coverage is required, sensors must be placed near to the ramp entrance; high coverage can be obtained at the positions according to the distributions obtained shown in Figure 2.22.

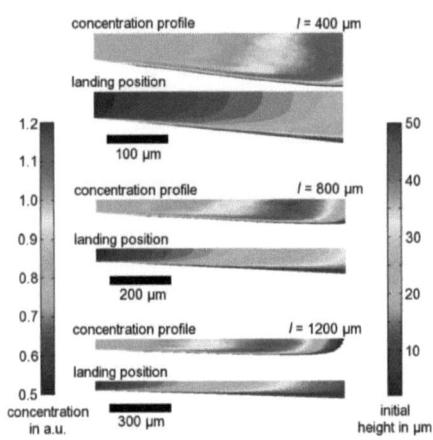

Figure 2.22: Results of the numerical calculations showing the concentration profiles, the landing position for different initial heights and the position of the concentration maximum for the parameters $h_1 = 50$ μm, $a_1 = 80$ μm, $h_2 = 20$ μm, $a_2 = 200$ μm and $u_{in} = 200$ μm/s. Each pair of graphics correspond to a different geometry length L. From the top to the bottom L is given by 400, 800 and 1200 μm. The upper plot shows the resulting concentration profile, the lower the projection of the initial height at the entrance along the vector field v.

Additionally, the lower subplot shows a projection of the initial height at the entrance cross section along the velocity field v given by (2.27). These data are obtained, by solving the Level-set-equation (1.19)

$$\frac{\partial \Phi}{\partial t} - D\Delta\Phi + \nabla(v\Phi) = 0 \qquad (1.19)$$

as introduced in section 1.3.1 by setting $\Phi = z$ along the channel entrance. Here, we chose $D = 0$ but solve (1.19) in a Petrov-Galerkin frame to maintain numerical stability. This plot gives information on the resulting concentration profiles if an arbitrary distribution is assumed instead of the constant Dirichlet condition $c = c_0$. Experimentally, this depends on the way the particles are diluted. To quantify the efficiency of the proposed microfluidic structure, the total capture rate of the inflowing material is analyzed, i.e. the percentage of the inflowing material that reaches the bottom of the geometry before leaving the ramp. The resulting values are compared to the corresponding rates of a straight channel of identical L, h_1 and w_1. The dependency of the capture rate on the parameters L, u_{in}, ξ and ρ_{part} are shown in Figure 2.23.

2.4 Particle positioning

The subplots (a), (c) and (d) indicate a strong increase of the particles reaching the bottom of the structure. The capture rate increases by up to more than 100 % in comparison to the straight channel geometry. Even for short distances high capture rates can be achieved (Figure 2.23(a)). Additionally, particles of high density ρ_{part} benefit even stronger (Figure 2.23(d)).

As incompressible liquids are discussed, the flow into the ramp equals the flow out of the ramp. Therefore, the time t_S an unbound particle remains in the ramp is given by the ratio of the ramp volume V and the inflow Γ. For constant h_2 the volume of the ramp is given by

$$V = \frac{L}{6} A_1 \left(2 + \frac{h_2}{h_1} + \chi \left(\frac{h_1}{h_2} + 2 \right) \right) \tag{2.28}$$

The inflow Γ is obtained by integrating the velocity profile u along the ramp inlet

$$\Gamma = \int_{\text{inlet}} \langle u, \hat{n} \rangle \, dr . \tag{2.29}$$

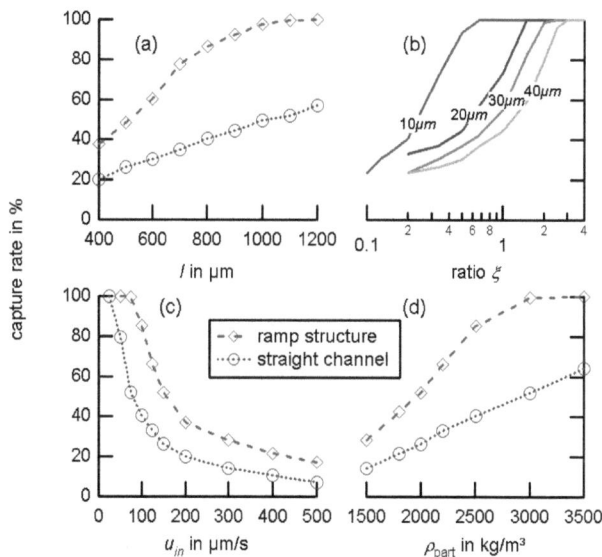

Figure 2.23: Calculated capture rates of the device in comparison to a straight channel for different lengths (a), cross section ratios (b), inflow velocities (c) and particle densities (d). If system coefficients are not explicitly given, it is $L = 800\ \mu m$, $\xi = 1$, $u_{in} = 200\ \mu m/s$ and $\rho_{part} = 2500\ kg/m^3$.

2. Microfluidic devices

Thus, the time t_s decreases if χ decreases. However, this does not necessarily lead to an acceleration of the capturing process as can be seen in Figure 2.23(b) since with decreasing χ the capture rate of 100 % cannot be maintained. In the cases of Figure 2.23(b), the minimal t_s for a 100 % capture rate is given by 8.4 s ($h_2 = 40$ μm), 6.6 s ($h_2 = 30$ μm), 6.1 s ($h_2 = 20$ μm) and 4.3 s ($h_2 = 10$ μm). Generally, a further decrease of t_s can be achieved by a further flattening of the exit. However, limitations are given by the experimental realization of the fluidic structure. Furthermore, as long as a constant capture rate is ensured doubling the velocity scale results in halving the capture time scale. The ramp structure enables a 100 % rate at far higher velocities than the straight channel, Figure 2.23(c), thus decreasing the capture time as well.

At this point, we may conclude that the proposed ramp structure enables a fast positioning of particles along a specified target region. Since the device only bases on hydrodynamic and gravitational effects, a low degree of complexity is obtained making it easy to implement such a structure into existing microfluidic devices. However, though not necessarily apparent, the very simple setup is bought at a cost of geometrical complexity. The three-dimensional ramp structure can no longer be realized by standard lithography methods. Instead, the channel structure was realized using an injection molding technique provided by Reiner, 2008, the applied mold mask was fabricated by a milling process. Since typical sensor arrays can be on a size scale of up to several millimeters (Edelstein et al., 2000) (~ 2 mm in the framework of the MrBead-project), the parameters used for the experimental realization differ slightly from the theoretical considerations. For the actual setup, it is $h_1 = 50$ μm, $a_1 = 80$ μm, $h_2 = 25$ μm, $a_2 = 300$ μm and $L = 3.3$ mm are chosen to make the device applicable to a wide range of existing sensor arrays. Deviating from the theoretical calculations a longer ramp segment is analyzed here. Nevertheless, a tendency can be deduced from the simulation results. For the experimental observations carried out by F. Wittbracht, 2009, the same setup as in the previous sections was used. To realize the flow through the channels an oxygen plasma treatment of the polycarbonate plates with implemented channel is carried out leading to a reduction in the contact angle. Pressure-driven-flow through channels is achieved by hydrostatic pressure applying a 1 μl droplet to the channel entrance leading to an inflow velocity at the entrance of the ramp of about 70 μm/s. The velocity is thereby determined by the bead velocity. The binding of a bead to the bottom plate can be detected optically in situ.

In the framework of the experiments, different bead solutions were investigated. Besides biotin-coated and uncoated Chemagen beads also Dynabeads MyOne^{TM} with

2.4 Particle positioning

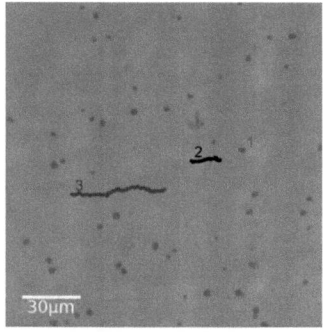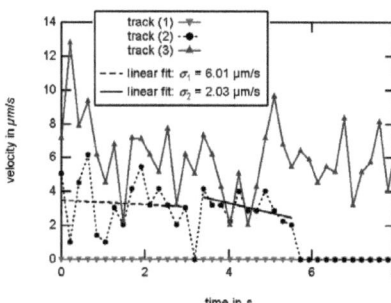

Figure 2.24: Experimental results. (a) shows optical microscopy image of typical bead tracks. Three different track types can be observed: (1) An immobilized bead shows no change in position over the whole time series. (2) Beads binding to the surface show an abrupt vanishing velocity. (3) Non-immobilized beads follow the velocity profile of the liquid where their velocity by Brownian motion. (b) presents a time-dependent velocimetry of the of the presented beads tracks of (a). Fitting track (2) by a linear function along the time interval [0 s, 3.25 s] and [3.5 s, 5.5 s] reveals a decrease of the diffusion constant to $\sigma_1/\sigma_2 \approx 1/3$.

a carboxylic acid functionalization are used as a reference. The concentration of all bead solutions is adjusted to $c = 0.1$ mg/ml. Dilutions are realized with DI-water and a PBS-buffer provided by MicroCoat, 2008. Typical bead tracks as observed in the experiment are shown in Fig. 2.24(b). Three different states of movement can be identified: Beads that are already bound to the bottom plate show no velocity (1). A binding event can be observed due to the spontaneous vanishing velocity (2). Beads that show no immobilization are characterized by a non-vanishing velocity (3). Immobilization of beads depends on the surface functionalization of beads and bottom plate and the diluting agent. The immobilization behaviour of the different bead species is summarized in Table 2.2. Uncoated Chemagen beads show no immobilization, if they are diluted with DI-water, however, they immobilize if they are diluted with PBS-buffer. This effect is caused by the surface polarity of the beads and bottom plate with respect to the buffer polarity. The biotinilated Chemagen beads show immobilization in the DI-water and PBS-buffer dilutions. Analyzing the MyOne beads leads to the observation of no immobilization in the reference system. For further verification a mixture of biotinilated Chemagen beads and MyOne beads is analyzed.

Both diluting agents are used. Immobilization can only be detected for the Chemagen but not for the MyOne beads. The immobilization of beads, if a ligand-receptor pair is present, is a clear indication of beads reaching the channel ground. Brownian

2. Microfluidic devices

Table 2.2: Immobilization behaviour of different bead species and functionalizations used in the experiments.

	PBS buffer	DI water
MyOne™ (COOH)	no immobilization	no immobilization
Chemagen (biotin)	immobilization	---
Chemagen (uncoated)	immobilization	no immobilization
mixture:		
MyOne™ (COOH)	no immobilization	no immobilization
Chemagen (biotin)	immobilization	immobilization

motion decreases directly before the immobilization event, as shown in Figure 2.24(b). This can be attributed to the reduction of diffusion in the vicinity of walls. The decrease of the diffusion constant D can be quantified by fitting the time intervals [0 s; 3.25 s] and [3.5 s; 5.5 s] in the case (2) track in Figure 2.24(b) independently by linear functions. We find standard deviations σ of the velocity with a relation of 1/3, which is in very good agreement with the theoretical prediction and the findings reported (Faucheux and Libchaber, 1994).

Evaluating the surface coverage of immobilized beads on the bottom plate at the end of the ramp section with respect to the position in x-direction leads to a surface coverage distribution presented in Figure 2.25. A homogenous coverage of the bottom plate with immobilized beads can be found. This coincides with the numerical results (Figure 2.22) where a homogeneous, enhanced (red area of the concentration plots) concentration value close to the channel exit is obtained. However, extrapolating the numerical results leads to the expectation that the maximum should be found in some distance in front of the exit. At the distance chosen for the experimental observations all particles should already be bound to the functionalized surface. One possible reason for these devia

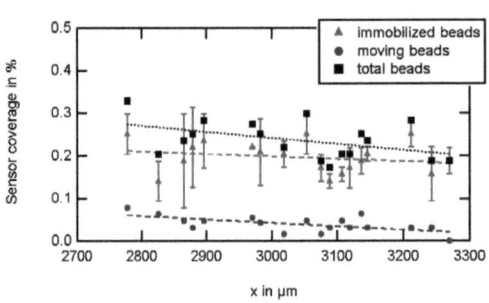

Figure 2.25: Surface coverage distribution of beads on the channel ground of the microfluidic structure.

tions can be given as follows: As already mentioned, the boundary conditions chosen imply an instantaneous binding of the particles to the ground. Probably this does not hold in the experimental situation, here particles will be dragged along the functionalized surface until the binding takes place. The concentration profile is therefore shifted towards the channel exit in the experimental case.

2.5 Conclusions

We presented a μTAS-system for preparation/reaction, separation and detection tasks. The theoretical design guidelines proposed in this work have been verified by the experiments of F. Wittbracht, B. Eickenberg and A. Auge, and overall a very strong agreement of the predictions from finite element simulations and experimental observations has been reported.

The microfluidic separation device presented in section 2.2 enables the separation of magnetic beads by using hydrodynamic and magnetic forces. We theoretically predicted the behaviour of the system and experimentally proved that it is suitable for the separation of magnetic beads with respect to size. Therefore, employing magnetic particles of different magnetic properties and choosing different surface functionalization, this device is suitable for the separation of different biomolecules (e.g. antibodies) by magnetic carriers. Furthermore, the device has been designed to suppress diffusive effects, enabling separation of particles on the nanoscale. It was shown that the separation device maintains its functionality even for particles down to the size scale of several 10s of nanometers, despite a strong diffusive motion contribution if a high enough velocity ratio ξ in both channels is created. We found that for a given property e.g. size the parameter ξ determines which values of this property can still be separated from each other. Therefore, the degree of adjustment of ξ is important for the adaptability to different particle species and is thus a crucial requirement for total lab-on-a-chip systems.

The integration of the separation device in a lab-on-a-chip structure was analyzed in section 2.3. The adjustment of important hydrodynamic flow properties (velocity ratio, beam position in separation site) could be achieved by variation of a small number of geometry parameters. It was numerically shown that transport and separation properties are marginally affected by particle diffusion down to a particle size scale of 10 nm. To maintain stability we theoretically predicted an increase of the inflow velocity leading to a geometry limitation law connecting duration time in the

2. Microfluidic devices

reaction site which may be necessary to allow chemical reactions to take place and particle size linearly. Due to the design of the channel structure all tasks were realized without additional microfluidic components e.g. micropumps of microvalves. The design is therefore easy to integrate and might help to reduce the complexity of existing lab-on-a-chip devices.

For the positioning of the magnetic carriers a microfluidic ramp structure was discussed in section 2.4 employing only gravity. The device thereby shows a very homogenous concentration over a long range, as well as a local maximum enabling different types of applications. We experimentally proved the depositing of magnetic beads on the bottom of the device using surface coatings of beads and bottom plates. The evaluation of the surface coverage shows a homogenous distribution close to the channel exit which also can be found in the theoretical prediction. The ramp structure provides high yield at small time scales. We have also proven by numerical simulations that the device can be used as positioning system for particles in the flow. For this application particles have to be provided at a certain channel height. The landing position follows from the proposed model and predicts a narrow spatial scattering.

At this point only the experimental proof for the combined system is still missing which is work in progress and the focus of current Bachelor and Master theses.

2.5.1 Outlook

An idea to further simplify the device is to combine the separation and the positioning devices. It is apparent that the proposed ramp geometry from section 2.4 can easily be extended to a separation device of differently functionalized particles. Therefore, the parameters have to be adjusted so that a capture rate of 100% is achieved. All particles reach the bottom of the geometry and will thus bind on a functionalized surface if they are correctly coated. Certain particle species will completely be filtered from the dilution. The separation will occur on the above mentioned time scale and is therefore also very fast.

In this sense, it might be interesting to note the following observation when evaluating the results of the 'Level set'-approach for the calculation of the projection of initial heights along the convection field. If we divide the entrance height into equidistant height segments, we can calculate the projected area of each individual segment at the bottom of the ramp structure; the result is shown in Figure 2.26. The plots show a local maximum followed by local minimum which is in agreement with concentration profiles presented in Figure 2.22 corresponding to the inverse area. Simulations predict an area distribution which follows some sort of master curve W independent of the geometrical parameters. W completely characterizes the device. The capture rate can be readily obtained via the integral

$$\int_0^L W(x)dx = \frac{L}{2}(a_1 + a_2) \qquad (2.30)$$

The master curve shows a functional dependence in respect to the particle properties. Placing sensors for certain particle species at the minimum of W enables the detection of several antibodies in a single chamber, an additional separation device would no longer be necessary.

However, we have not found a direct way to calculate W for a given set of parameter which would be very helpful to verify the independence of geometrical and material parameters. Also from the experimental point of view more observation data are necessary of verify the existence of such behaviour.

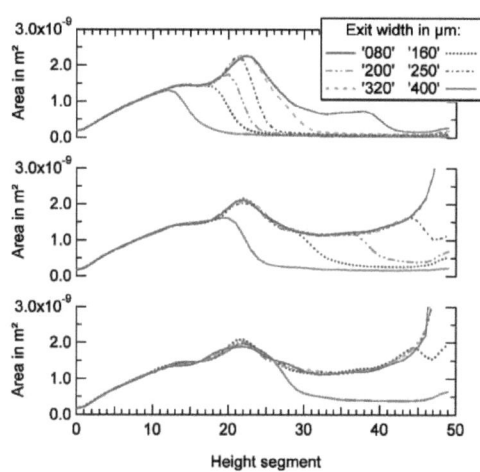

Figure 2.26: Master curve for different ramp lengths, from top to bottom 400, 800, 1200 µm.

2. Microfluidic devices

Chapter 3

Micromagnetism

When dealing with hydrodynamic systems, the continuum hypothesis proves key; it allows for the possibility of the separation of different scales and, in so doing, permits one to apply an effective, mesoscopic theory. In principal, this is a general assumption whenever different scales must be bridged, microscopic details lead to a macro- / mesoscopic behaviour. The details themselves are no longer important, but enter into the continuous model via a set of phenomenological material dependent parameters e.g. density and viscosity in the case of hydrodynamics. A rather similar approach can be employed for the description of magnetic materials. On the microscopic level magnetism arises due to the spins of atoms and their coupling. We may therefore model a solid by a lattice of spins as shown in Figure 3.1. Depending on the coupling between adjacent spins different behaviour may be found. In this thesis, we will mainly be interested in ferromagnetic materials (Landau and Lifshitz, 1935) which can be described by the theory of *micromagnetism*. The governing equations may be derived in a variational formalism: The equilibrium state leads to a minimum of the total free energy

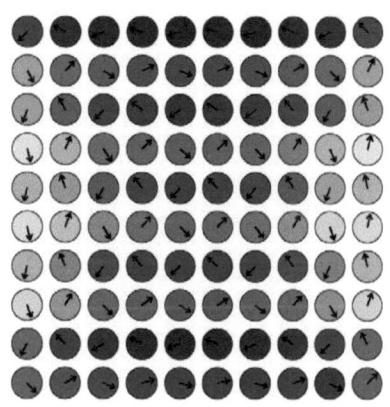

Figure 3.1: Modelling a magnetic solid by a spin grid, with the spin position at the grid nodes.

69

3. Micromagnetism

of the system. This energy may be decomposed into several contributions which will be introduced in sections 3.1 to 3.3. In particular, similar to our introduction of the theory of hydrodynamics, we explain how to bridge the gap between microscopic details and mesoscopic behaviour. In section 3.4, we introduce the *Brown equation* as the governing equation of static micromagnetism and the *Landau-Lifshitz/Landau-Lifshitz-Gilbert equation* in 3.5 as the respective equation for the phenomenological analysis of time-dependent phenomena.

3.1 From atomic to mesoscopic magnetism

On the microscopic level, we can consider solids to be perfectly periodic assemblies of atoms. To understand the magnetic behaviour of such systems, we assume each atom at a position given by R_i to carry a spin S_i (compare Figure 3.1). The sign of the *exchange integral J* determines whether neighbouring spins tend to align to each other ($J > 0$ – *ferromagnetic coupling*) or prefer an antiparallel orientation ($J < 0$ – *antiferromagnetic coupling*). The value of J depends on the degree of confinement of the atomic orbitals. To investigate the behaviour of magnetic systems, the Hamiltonian spin operator

$$\mathcal{H} = -\sum_{i \neq j} J(R_i, R_j) S_i S_j \qquad (3.1)$$

can be employed. It is well suited for the description of ordered ferro- and anitferromagnetic materials, spin waves as well as spin glasses, but may not be applied for the description of itinerate magnetic systems since in these cases the magnetism arises from delocalized electrons. In the special case of a cubic lattice symmetry (3.1) simplifies to

$$\mathcal{H} = -2J \sum_{i \neq j} S_i S_j \qquad (3.2)$$

with the sum expanded only over adjacent spins. Similar to our considerations when motivating the effective (continuum) theory of hydrodynamics, magnetic solids on the size scale of up to several micrometers consist of too large a number of spins to treat them individually. Focusing on the case of a strong ferromagnetic coupling be-

3.1 Atomic magnetism

tween adjacent spins $\hat{m}_i$ and $\hat{m}_j$ and writing $S_i = S\hat{m}_i$, we may assume that the force between adjacent spins is strong enough to only allow slight angle variations $\theta_{ij} = \sphericalangle(\hat{m}_i, \hat{m}_j)$. Therefore, the approximation $|\theta_{ij}| = |\hat{m}_i - \hat{m}_j|$ holds and (3.2) can be rewritten according to

$$\mathcal{H} = -2J \sum_{i \neq j} S_i S_j = -2JS^2 \sum_{i \neq j} \cos\theta_{ij} \approx -2JS \sum_{i \neq j}(1 - \tfrac{1}{2}\theta_{ij}^2)$$
$$= \text{const} + JS^2 \sum_{i \neq j} \cos\theta_{ij}^2 = \text{const} + JS^2 \sum_{i \neq j}(\hat{m}_j - \hat{m}_i)^2 \tag{3.3}$$

At this point, we state the existence of a continuous function $\hat{m}$ satisfying

$$|\hat{m}_j - \hat{m}_i| = \langle \Delta r_j, \nabla \hat{m} \rangle \tag{3.4}$$

with $\Delta r_j = R_j - R_i$. By introducing (3.4), the discrete spin dynamics (3.2) can then be treated in a continuous theory. Therefore, instead of obtaining energy contributions per spin, an energy density e_A may be calculated denoting the energy per unit volume. Decomposing the function $\hat{m}$ into its components, we obtain

$$\mathcal{H} = \text{const} + JS^2 \sum_{i \neq j}(\Delta r_j \cdot \nabla \hat{m})^2$$
$$= \text{const} + JS^2 \sum_{i \neq j}\left((\Delta x_j \cdot \nabla \hat{m}_x)^2 + (\Delta y_j \cdot \nabla \hat{m}_y)^2 + (\Delta z_j \cdot \nabla \hat{m}_z)^2\right)$$

Since $\sum_j \Delta x_j \Delta y_j = 0$ and $\sum_j \Delta x_j^2 = \frac{1}{3}\sum_j \Delta r_j^2$

are valid for crystals of cubic symmetry, it is

$$e_A = A\left((\nabla m_x)^2 + (\nabla m_y)^2 + (\nabla m_z)^2\right) \quad \text{with} \quad A = \frac{NJS^2}{6}\sum_j \Delta r_j^2 \tag{3.5}$$

and N the number of spins per unit volume.

The *exchange constant A* is a measure of the "magnetic stiffness" of a material. High values indicate a strong coupling between neighbouring spins, which will overcome external influences to a certain degree. Typical values range from 10^{-12} to sev-

3. Micromagnetism

eral 10^{-11} J/m. A systematic analysis of the stiffness term can be found in Döring's review of micromagnetics (Döring, 1966) where the generalized expression

$$e_A = \sum_{i,j,k} A_{jk} \frac{\partial \hat{m}_i}{\partial x_j} \frac{\partial \hat{m}_i}{\partial x_k}$$

is derived. Here, A is a symmetric tensor degenerating to a scalar for cubic or isotropic materials. In principle, hexagonal or other lower symmetry crystals require more exchange stiffness constants. However, in practice, the isotropic formula is used in all calculations; no experimental determination of anisotropic exchange stiffness coefficients has been recorded (Hubert and Schäfer, 2000).

3.2 Coupling between mesoscopic and atomic structure

On the microscopic level, the structure of a crystal is given (if we, for the moment, disregard defects such as vacancies and dislocations etc.) by a periodic assembly of a certain elementary cell; different crystals can be classified by their symmetry group (Bradley and Cracknell, 1972). Due to spin-orbit coupling, this (periodic) substructure induces a direction dependent energy, i.e. the *magnetocrystalline anisotropy energy* of the undisturbed crystal. Additional *induced anisotropies* arise from deviations from the perfect crystalline symmetry which may be introduced by e.g. lattice defects. Since the anisotropy energy needs to maintain the symmetry of the crystal, the dependency is expanded in a series of spherical harmonics where in most cases only the first (or occasionally the first two) contributions are considered since thermal agitation of the spins tend to average out the higher-order terms.

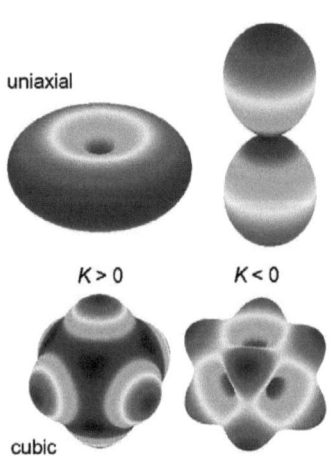

Figure 3.2: Energy surfaces of magnetic anisotropy energy considering only first order contributions; blue coincide with the energetically favourable orientations. In the uniaxial case for $K < 0$, the easy direction goes over to an *easy plane* with normal $\hat{e}$.

Figure 3.2 shows energy surfaces for

3.2 Magnetic anisotropy

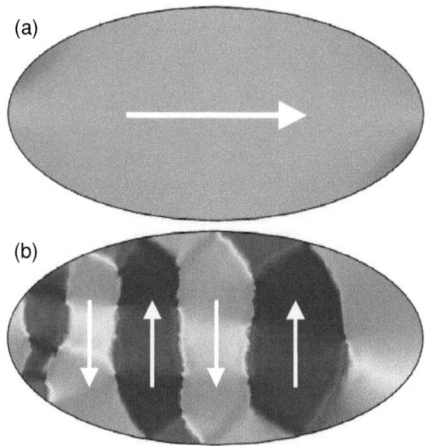

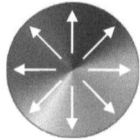

Figure 3.3: Magnetization configurations of an elliptical geometry, the colour code corresponds to the angle of the magnetization orientation. (a) No anisotropy is considered, the magnetization aligns with the geometrical easy axis due to stray field minimization. (b) A uniaxial anisotropy with easy axis parallel to the short semiaxis is assumed. The orientation aligns with the anisotropy direction, stray field energy though leads to the creation of closed magnetization "loops".

two very important examples: uniaxial and cubic symmetries. Energetically favourable orientations correspond to blue areas, whereas energetically unfavourable ones correspond to the red. The directions of minimum and maximum energies are referred to as *easy* and *hard directions*, respectively. In formula, they are given as follows:

a) The upper plots of Figure 3.2 correspond to a *uniaxial anisotropy*. The energy contribution depends on the relative orientation between the vector $\hat{m}$ and the "crystal direction" $\hat{e}$. Up to its fourth-order terms, the anisotropy energy density is given by

$$e_{\text{uni}} = K_1^{\text{uni}} \langle \hat{m}, \hat{e} \rangle^2 + K_2^{\text{uni}} \langle \hat{m}, \hat{e} \rangle^4 \tag{3.6}$$

The actual type of anisotropy depends on the values of K_1^{uni}. If K_1^{uni} is positive and large in respect to K_2^{uni}, an *easy axis* is obtained. For large negative K_1^{uni}, we find an *easy plane* or *planar anisotropy*. Intermediate values $0 > K_1^{\text{uni}} / K_2^{\text{uni}} > -2$ lead to an easy directions on a cone with angle θ relative to the axis given by $\sin^2 \theta = -\frac{1}{2} K_1^{\text{uni}} / K_2^{\text{uni}}$. This situation is a *conical anisotropy* (Hubert and Schäfer, 2000).

b) The lower plots of Figure 3.2 show *cubic anisotropies*. The general formula is given by

$$e_{\text{cub}} = K_1^{\text{cub}} \left(m_x^2 m_y^2 + m_x^2 m_z^2 + m_y^2 m_z^2 \right) + K_2^{\text{cub}} m_x^2 m_y^2 m_z^2 \tag{3.7}$$

73

3. Micromagnetism

denoting by m_i the magnetization components along the cubic axes. The material constants K_2^{cub} and higher-order terms can mostly be neglected. The constant K_1^{cub} assumes values in the range of $\pm 10^4$ J/m^3 for different materials. The sign of K_1^{cub} determines whether the $\langle 100 \rangle$ or the $\langle 111 \rangle$ directions are the easy directions for the magnetization (compare Figure 3.2).

It should be noted that uniaxial anisotropies can be much stronger than cubic anisotropies, reaching some 10^7 J/m^3 for rare earth transition metal permanent magnetic materials.

3.3 Magnetostatics in matter

A magnetic material of a spatial magnetization distribution M creates a magnetic field H in all of space which can be calculated by solving the Maxwell equations for matter. If no external current densities are considered, the PDE system is given by the homogeneous set of equations for magnetic field H and magnetic flux density B

$$\nabla B = 0 \quad (3.8a) \qquad \nabla \times H = 0 \quad (3.8b)$$

Due to (3.8b) the magnetic field H may be written in the form $H = -\nabla \phi_{mag}$ with a scalar potential ϕ_{mag}. Further, field and flux density are connected via

$$B = \mu_0 (M + H) = \mu_0 (M - \nabla \phi_{mag}). \quad (3.9)$$

Combining equations (3.9) and (3.8a), we retrieve the inhomogeneous Laplace equation for the magnetic potential ϕ_{mag}

$$\Delta \phi_{mag} = \nabla M.$$

In section 3.2, magnetocrystalline anisotropy effects were discussed. A different type of anisotropy may be induced by the shape of the magnetic domain Ω_{mag}. Figure 3.3(a) shows the magnetic configuration of an ellipse in the absence of magnetocrystalline anisotropy contributions: the orientation aligns with the long axis. The origin of this orientation is, however, of a completely different nature and coupled to the energy E_{stray} of the magnetic field introduced by the magnetic material in the surrounding space. In general, it is

$$E_{\text{stray}} = \frac{\mu_0}{2}\int_{\mathbb{R}^3} H^2\, dr = \frac{\mu_0}{2}\int_{\mathbb{R}^3}\left\langle H, \left(\frac{B}{\mu_0} - M\right)\right\rangle dr = -\frac{\mu_0}{2}\int_{\Omega_{\text{mag}}}\langle H, M\rangle\, dr \equiv \int_{\Omega_{\text{mag}}} e_{\text{demag}}\, dr.$$

with $\quad e_{\text{demag}} = -\dfrac{\mu_0}{2}\langle H, M\rangle \qquad\qquad$ (3.10)

this energy is often referred to as *demagnetization energy*. Considering an additional crystal anisotropy, a complex interplay between different energy contributions may arise which leads to complex phenomena. Figure 3.3(b) shows the situation of Figure 3.3(a) with an additional uniaxial anisotropy with easy axis along the y-axis. Since a y-orientated magnetic distribution would entail large stray field energy, an array of antiparallel areas can be found. Each subdomain aligns with the easy crystal axis while the antiparallel orientation of adjacent *domains* minimizes the stray field contribution.

Finally, if a magnetic volume is brought into an external field H_{ext}, its energy density is given by the *Zeeman energy*

$$e_{\text{Zeeman}} = -\mu_0 \langle M, H_{\text{ext}}\rangle. \qquad\qquad (3.11)$$

3.4 Static micromagnetism

The energy contributions explained in the preliminary sections allow for the determination of the equilibrium state of the magnetization distribution by means of variational calculus. The effective equations describing the behaviour of a magnetic material follow one of the most general physical principals: minimization of the total free energy E. According to the preliminary sections, we may therefore write

$$E = \int_{\Omega_{\text{mag}}} (e_A + e_{\text{ani}} + e_{\text{demag}} + e_{\text{Zeeman}})\, dr \qquad\qquad (3.12)$$

$$= \int_{\Omega_{\text{mag}}} \left(A\left((\nabla m_x)^2 + (\nabla m_y)^2 + (\nabla m_z)^2\right) + e_{\text{ani}} - \frac{\mu_0}{2}\langle H, M\rangle - \mu_0\langle M, H_{\text{ext}}\rangle\right) dr$$

3. Micromagnetism

with e_{ani} a function describing an arbitrary orientation dependence. In order to find the equilibrium state, the integral expression must be minimized under the constraint $|\hat{m}|=1$. To maintain this condition, an additional Lagrange parameter λ_m is introduced. Finally, writing $\boldsymbol{H} = -\nabla \phi_{mag}$, we obtain

$$\tilde{E}[\hat{m}, \phi_{mag}, \lambda_m] = \int_{\Omega_{mag}} \left(A\left((\nabla m_x)^2 + (\nabla m_y)^2 + (\nabla m_z)^2\right) + e_{ani} \right. \qquad (3.13)$$

$$\left. + \frac{\mu_0}{2} \langle \nabla \phi_{mag}, \boldsymbol{M} \rangle - \mu_0 \langle \boldsymbol{M}, \boldsymbol{H}_{ext} \rangle \right) d\boldsymbol{r} + \int_{\Omega_{mag}} \lambda_m (m_x^2 + m_y^2 + m_z^2 - 1) d\boldsymbol{r}$$

The equilibrium state is given by the angle distribution $\hat{m}$ which makes $\tilde{E}$ stationary, i.e. it needs to satisfy

$$\frac{\delta \tilde{E}[\hat{m}, \phi_{mag}, \lambda_m]}{\delta \hat{m}} = 0 \qquad (3.14)$$

The simultaneous variation of all energy contributions can be found in Appendix A.3. The result is given by the *Brown equation*

$$\hat{m} \times \boldsymbol{H}_{eff} = 0 \qquad \text{on } \Omega$$

$$\text{with} \quad \boldsymbol{H}_{eff} = -\frac{2A}{\mu_0 M_s} \Delta \hat{m} - \frac{\delta e_{ani}(\hat{m})}{\delta \hat{m}} - \nabla \phi_{mag} + \boldsymbol{H}_{ex} \quad \text{on } \Omega \qquad (3.15)$$

$$\text{and} \quad |\hat{m}| = 1 \qquad \text{on } \partial\Omega$$

For the boundary conditions, a homogeneous Neumann condition may be employed for the description of a free magnetic surface. In many applications though, this assumption is not correct. Thin magnetic films obtain an alignment perpendicular to the layer plane due a strong *surface anisotropy* which has been e.g. in granular CoZrO-systems (Sun et al., 2005). Such configurations have promising applications for data storage devices as the perpendicular magnetization orientation may increase the data / area-ratio. Magnetic multi-layer systems are another common example where homogeneous Neumann conditions do not resemble the proposed situation:

3.4 Static micromagnetism

several magnetic films are coupled to each other across a non-magnetic layer (e.g by RKKY-coupling, *GMR-junctions* (Grünberg et al., 1986; Baibich et al., 1988) or by tunnelling processes, *TMR-junctions*, see chapter 5). In the most general formulation, the behaviour along the boundary of a magnetic material with magnetization direction $\hat{m}$ in contact with a second magnetic material described by $\hat{m}'$ is given by (Labrune and Miltat, 1995)

$$\hat{m} \times \left(2A(\hat{n}\nabla)\hat{m} + \frac{\delta e_{\text{surf}}(\hat{m},\hat{m}')}{\delta \hat{m}} - (C_{\text{bl}} + 2C_{\text{bq}}\langle \hat{m},\hat{m}' \rangle)\hat{m}' \right) = 0. \quad (3.16)$$

e_{surf} denotes a (surface) anisotropy energy density while C_{bl} and C_{bq} are the *bilinear* and *biquadratic coupling parameters*. Examples for layer coupling can be found in the systems discussed in chapter 5.

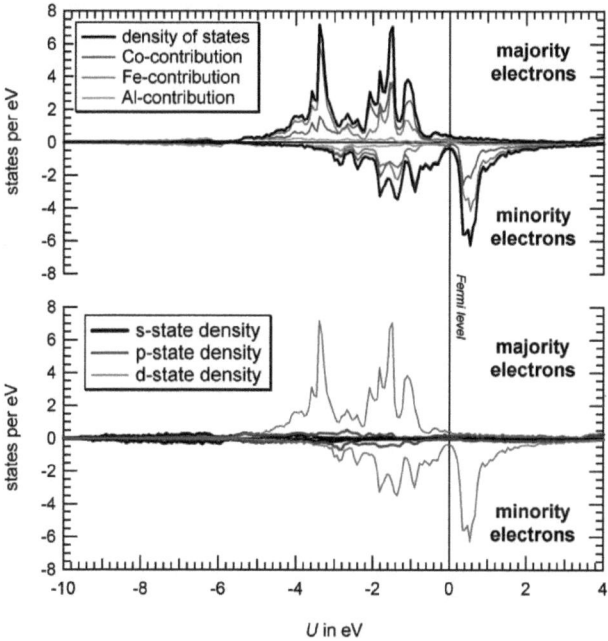

Figure 3.4: Density of states for a Co_2FeAl crystal assembled in the L_21-lattice structure, i.e. for interpenetrating fcc-lattices with base positions at $(0\,0\,0)$, $(\frac{1}{4}\,\frac{1}{4}\,\frac{1}{4})$, $(\frac{1}{2}\,\frac{1}{2}\,\frac{1}{2})$ and $(\frac{3}{4}\,\frac{3}{4}\,\frac{3}{4})$. Due to asymmetric behaviour of spin-up and spin-down electrons, the material possesses a non-zero magnetic moment $\sim 4\,\mu_B$ with μ_B the *Bohr magnetron*. The lower plot shows the contribution of different electrons.

3. Micromagnetism

Equation (3.15) contains the most common contributions considered in micromagnetic calculations. Possible extensions are e.g. the incorporation of external stresses or magnetostriction both adding an additional term to the effective magnetic field via the variational derivative of their energy densities (see e.g. Hubert and Schäfer, 2000). Similar to the Navier-Stokes equation, the Brown equation contains several material coefficients which arise due to the micro-details. The origin of the exchange constant A (arising because of the coupling between adjacent atoms) was already discussed in section 3.1. The saturation magnetization, on the other hand, is a measure for "unsymmetric" spin orientation distribution in the unit cell. In the case of a magnetic material, a certain spin orientation dominates. Information on the magnetic structure can be obtained via *band structure calculations* based on the *density functional theory* (DFT) (Gonis and Butler, 2000). An example for a Co_2FeAl-crystal in the $L2_1$-structure calculated by the SPRKKR-package by H. Evert (SPRKKR, 2006) is shown in Figure 3.4.

3.5 Dynamic micromagnetism

Equation (3.15) may be used to calculate the equilibrium state of a magnetization distribution. However, it does not give any information on how this state was reached. If a magnetic moment is brought into an (effective) external magnetic field, it starts to precess around the field direction with the *Lamor frequency* $\omega_{Lamor} = \gamma | H_{eff} |$ with the *gyromagnetic ratio* γ. The dynamic behaviour is given according to

$$\frac{d\hat{m}}{dt} = -\gamma \hat{m} \times H_{eff} \quad \text{with} \quad \gamma = \frac{\mu_0 g e}{2 m_{elec}} = g \cdot 1.105 \cdot 10^5 \frac{m}{As} \quad (3.17)$$

with e the elementary charge, m_{elec} the electron mass and g the *Landé factor* which is for many ferromagnetic materials given by $g \approx 2$. The dynamics obtained from (3.17) describe a precession of the magnetic vector $\hat{m}$ around the effective field H_{eff}. The angle enclosed does *not* change in respect to time (Figure 3.5) which occurs since no damping mechanisms have been taken into account. Damping of the precession originates from many different phenomena: eddy currents, macroscopic discontinuities (*Barkhausen jumps*), diffusion and the reorientation of lattice defects, or spin-scattering mechanisms can all introduce irreversibilities and losses. The first two ex-

3.5 Dynamic micromagnetism

amples introduce long-range dynamics which cannot be separated from the domain structure. However, all local effects can be summarized in a single term with a phenomenological, dimensionless damping parameter α describing the intrinsic loss. The dynamics are given by the *Landau-Lifshitz* equation (Landau and Lifshitz, 1935)

$$\frac{d\hat{m}}{dt} = -\gamma \hat{m} \times H_{\text{eff}} - \alpha \hat{m} \times \frac{d\hat{m}}{dt} \quad (3.18)$$

or alternatively, by substituting (3.18) into itself and reformulating (Gilbert, 1955)

$$\frac{d\hat{m}}{dt} = -\frac{\gamma}{1+\alpha^2} \hat{m} \times H_{\text{eff}} + \frac{\alpha\gamma}{1+\alpha^2} \hat{m} \times \frac{d\hat{m}}{dt}$$

by the *Landau-Lifshitz-Gilbert* equation. The parameter α can usually be found on a scale from 10^{-3} to 0.1. Different influences have been studied in the works of A. Azevedo, e.g. (Azevedo et al., 2000).

In order to investigate the influence of *spin currents*, an additional term needs to be added to (3.18) which can be found in the works of J. Slonczewski, 1996, and L. Berger, 1996.

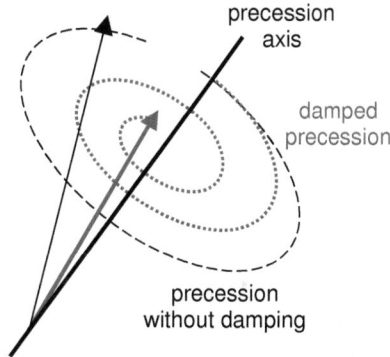

Figure 3.5: Precession without (dashed) and with (dotted) damping. The angle between magnetic vector and precession axis is constant over time if no damping is taken into account.

3. Micromagnetism

Chapter 4

Magnetically interacting particles

The Landau-Lifshitz equation describes the equilibrium state of a magnetization distribution for magnetic objects on the mesoscale. In particular, it holds for magnetic particles down to a particles size of several nanometers. However, if objects below a certain size scale are considered, magnetic domains can no longer be found. This is due to the fact that on small dimensions the exchange energy dominates all other contribution and does not allow for a spatial change of the magnetization distribution. This leads to the concept of *superparamagnetism* which will be explained in section 4.1; the properties of a superparamagnetic and consequently homogeneously magnetized sphere are summarized in section 4.2. Similar to paramagnetic materials, superparamagnetic objects show no stray field if no external field is applied. The particles employed for theoretical calculations and experimental observations in chapter 2 were assumed to be superparamagnetic. However, due to their sizes of about 1 to several micrometers they do not fulfil the size requirements, but their magnetic configuration should show a domain structure. In section 4.3, we will generalize such magnetic microparticles by the concept of magnetic beads which consist of superparamagnetic nano-components embedded in a polymer matrix. If we apply the Landau-Lifshitz equation to such systems the original set of partial differential equations may be reduced to a set of ordinary differential equations. Due to this substructure, the magnetic behaviour of such magnetic beads is influenced by the dipolar coupling

4. Interacting particles

of the nanoparticles. An analysis on the deviations of the static properties from the Langevin-formalism was recently achieved by V. Schaller et al., 2009a, b, by means of Monte Carlo simulations. In section 4.4, we investigate the consequences on the dynamic relaxation processes which result from such nano(sub)structure. Thereby, we employ a similar approach that was also used by D. Laroze et al., 2008, to study the dynamic behaviour of two interacting dipoles. A key assumption for the systems discussed in chapter 2 was that sufficiently low particle concentrations are employed, so that particle-particle interactions can be neglected. From the experimental point of view such an assumption can always be easily met but it does not allow for fast, high-throughput applications. These problems have been addressed in many works e.g. Wiklund et al., 2006, or Pamme and Manz, 2004. In the last section of this chapter, we consider particle-particle interactions for particles dissolved in liquid when brought into a homogeneous, external magnetic field where they agglomerate in rod-like structures. Self-assembles particle arrays have recently been in used in the group of M.A.M. Gijs (Lacharme et al., 2008, 2009; Sivagnanam et al., 2009) for the development of sandwich immunoassays. However, particles are trapped in such devices. In continuous-flow applications, such interactions are considered as negative side effects which lead to different effective rates and may also lead to a decreased device yield (Mikkelsen et al., 2005b). In section 4.5, a novel method is presented to readily employ the particle-particle interaction for controlling the behaviour of assemblies within a fluid flow. We will show that the particle flow may be uncoupled from such liquid flow by a homogeneous field only, and can be guided by the relative orientation of (fluid) flow and field direction as long as changes are adiabatically.

4.1 Superparamagnetism

Due to the interplay between different energy contributions discussed in chapter 3, the magnetic material splits up into several magnetic domains. When dealing with systems of decreasing sizes, certain energy contributions dominate. In particular, the exchange energy does not allow for any spatial change of the magnetization orientation below certain dimensions; the system consists of only a single domain. In this case, the behaviour of the system is governed by the interplay of the demagnetization field, magnetocrystalline anisotropy and external perturbations. For special cases such as the *Stoner-particle* (spherical geometry), the switching behaviour can be discussed by analytic means, leading to the so-called *switching asteroids* (Stoner and Wohlfarth, 1948), Figure 4.1.

The three remaining contributions depend on the volume of the particle, while the thermal energy $k_B T$ is constant with k_B, the Boltzmann constant, and T, the absolute temperature. Below a certain critical size, thermal agitation may be sufficient to switch the magnetization configuration between different minima. If we consider spherical particles with cubic anisotropy, thermal energy overcomes the switching barrier below the *superparamagnetic size limit* D_{sp} which is given by

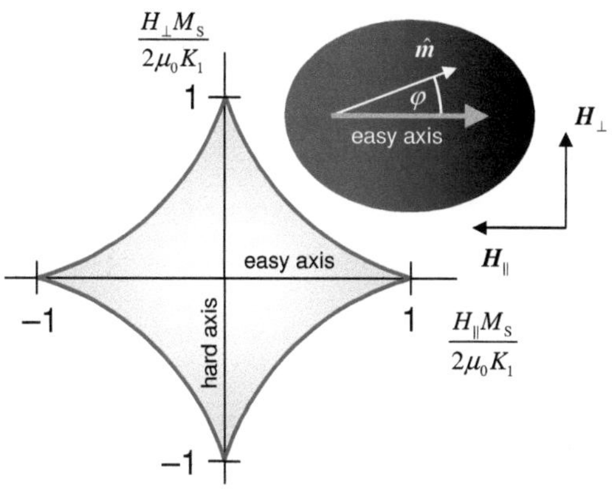

$$\frac{4}{3}\pi \left(\frac{D_{sp}}{2}\right)^3 = \frac{25 k_B T}{K_1} \quad (4.1)$$

with K_1 the first cubic anisotropy constant. Higher orders are not considered in this formula as the process is driven by thermal effects. The originally ferromagnetic material loses its memory: objects under the limit D_{sp} show no hysteresis but exert paramagnetic behaviour. Since their magnetic moments are far higher than paramagnetic objects, this is called *superparamagnetism*. Typical material values are given in Table 4.1.

Figure 4.1: Stoner-Wohlfarth asteroid. A single domain particle with uniaxial anisotropy (of arbitrary origin) changes its magnetization orientation if an external field is applied which lies outside the asteroid. In units of $H M_S/(2\mu_0 K_1)$ the switching lines are given by $-\cos^3\varphi$ and $\sin^3\varphi$ for parallel and orthogonal field, respectively.

Table 4.1: Magnetizations and superparamagnetic size limit for different materials [AHüt04].

	M_S in kA/m	D_{sp} in nm
bcc-Fe	1714	16.0
fcc-Co	1420	15.8
$Fe_{50}Co_{50}$	1910	23.6
Fe_3Co	1993	17.0
Fe_3O_4	415	28.0
Fe_2O_3	380	34.9

4. Interacting particles

4.2 Homogeneously magnetized spheres

In many model approaches investigating the influence of small magnetic particles, it is sufficient to model magnetic markers with a homogeneously magnetized sphere of radius R and magnetic moment $\boldsymbol{m}_{\text{part}}$. The stray field created by such a particle can be derived from the Maxwell equations of magnetostatics: $\nabla \boldsymbol{B} = 0$ and $\nabla \times \boldsymbol{H} = 0$. Due to the second equation, the magnetic field may be expressed by a scalar potential ϕ_{mag} in the form $\boldsymbol{H} = -\nabla \phi_{\text{mag}}$. Expressing the magnetic flux density by $\boldsymbol{B} = \mu_0(\boldsymbol{M} + \boldsymbol{H})$, ϕ_{mag} needs to satisfy the inhomogeneous Laplace equation

$$\Delta \phi_{\text{mag}} = \nabla \boldsymbol{M}. \qquad (4.1)$$

In the case of an unbounded domain, an analytic solution can only be found for highly symmetric systems. In our case, due to the axis-symmetry of the problem, an expansion of ϕ_{mag} into a series of the orthogonal basis set of the axissymmetric function space, the Legendre polynoms, is possible. Exploiting the infinity limits $|H| \to 0$ for $|r| \to \infty$ and the continuity of $H_{\parallel}$ and $B_{\perp}$ along material interfaces, it can be shown that (Jackson, 1975)

$$\boldsymbol{H}_{\text{part}}(\boldsymbol{r}) = \frac{1}{4\pi} \left(\frac{3 \langle \boldsymbol{m}_{\text{part}}, \Delta \boldsymbol{r} \rangle \Delta \boldsymbol{r}}{|\Delta \boldsymbol{r}|^5} - \frac{\boldsymbol{m}_{\text{part}}}{|\Delta \boldsymbol{r}|^3} \right) \qquad (4.2)$$

with $\Delta \boldsymbol{r} = \boldsymbol{r} - \boldsymbol{r}_{\text{part}}$ (compare Appendix A.4). A representation of the stray field by a streamline plot is shown in Figure 4.1(a). (b) and (c) show area cross-sections parallel to the xy- and the xz-plane under the assumption of $\boldsymbol{m}_{\text{part}} \parallel \hat{\boldsymbol{z}}$, respectively.

If a magnetic volume V of magnetization $\boldsymbol{M}$ is brought into an external field $\boldsymbol{H}$, it contains a potential energy density E depending on the relative orientation between $\boldsymbol{M}$ and $\boldsymbol{H}$. Consequently, a torque $\boldsymbol{\tau}$ results aligning the magnetization direction with the field orientation. Additionally, if an

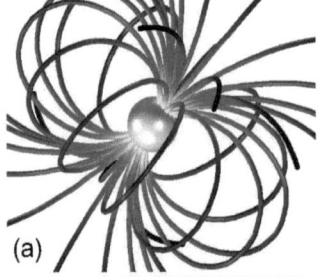
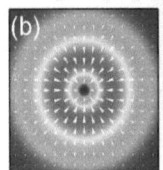
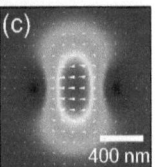

Figure 4.2: Stray field of a homogeneously magnetized sphere. (a) Streamline plot, (b), (c) in-plane components for $\boldsymbol{M}$ perpendicular and parallel to plane, respectively.

inhomogeneous field is considered, a magnetic force F_{mag} acts on the magnetic object

$$E = \mu_0 \int_V \langle M, H \rangle dr \qquad (4.3a)$$

$$\Rightarrow \quad F_{mag} = -\nabla E = \mu_0 \int_V (M\nabla) H \, dr \qquad (4.3b)$$

$$\text{and} \quad \tau = \int_V M \times H \, dr \qquad (4.3c)$$

It should be pointed out that a magnetic moment does not feel a force in a homogeneous field. This fact was already applied in the setup of the magnetic separation device discussed in chapter 2.2. A strong homogeneous field induced a torque aligning the moment vectors in z-direction, leading to a very specific particle motion.

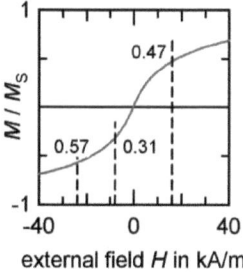

Figure 4.3: Magnetic behaviour of Dynabeads MyOne™ obtained from AGM-measurements. No hysteresis can be found.

4.3 Magnetization dynamics

When modelling magnetic particles in microfluidic devices a common assumption is a superparamagnetic behaviour. However, if we consider e.g. the Dynabeads MyOne™, the picture of a magnetically massive sphere cannot be correct since the particle dimensions can be found on a size scale of about 1 μm. The magnetic behaviour does not show any hysteresis though, as presented in Figure 4.3. Therefore, we need to refine the conception of a magnetic particle. In Figure 4.4 a schematic representation of the internal structure of a *magnetic bead* or *multi-core particle* is shown. Small superparamagnetic nanoparticles are embedded in a polymer

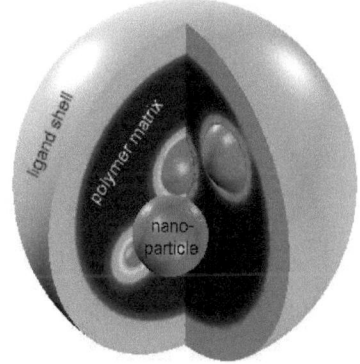

Figure 4.4: Internal structure of a magnetic bead. Superparamagnetic nanoparticles are embedded in polymer matrix. A protective ligand shell is employed to maintain stabilization.

85

4. Interacting particles

matrix that is stabilized by a ligand shell. The magnetic cores commonly consist of magnetite or maghemite, $M_s \approx 380$ kA/m, but also other materials of higher magnetization have been recently studied and promise higher responses to magnetic fields or stronger influence on detection devices. The response of free magnetic moments to an external homogeneous magnetic field is given by the *Langevin-function*

$$L(H) = \coth\left(\frac{\mu_0 \langle H, M \rangle}{k_B T}\right) - \frac{k_B T}{\mu_0 \langle H, M \rangle}. \tag{4.4}$$

Due to dipolar coupling between individual nanoobjects, the behaviour can differ slightly from (4.4). Deviations were analyzed by Schaller et al., 2009a, b, by means of Monte Carlo simulations. However, such an approach does not give information on the dynamic behaviour of such objects. To do so, we will analyze the system by solving the Landau-Lifshitz equation where we incorporate the simplifications introduced in sections 4.1 and 4.2.

Considering superparamagnetic components, the magnetization along each nanoparticle is constant in respect to space $\hat{m}(r,t) = \hat{m}(t)$. Therefore, the effective magnetic field H_{eff} introduced in (3.15) simplifies to

$$\boldsymbol{H}_{\text{eff}} = -\frac{\delta e_{\text{ani}}(\hat{m})}{\delta \hat{m}} - \nabla \phi_{\text{mag}} + \boldsymbol{H}_{\text{ex}}. \tag{4.5}$$

In particular, the dynamic equations no longer depend on space; the set of partial differential equations can be refigured as a set of ordinary equations. Considering a system of N such particles, the ODE system can be written in matrix form

$$(\text{Id} - \alpha \mathcal{M})\frac{\partial \tilde{m}}{\partial t} = \gamma \mathcal{M} \tilde{H}_{\text{eff}}, \tag{4.6}$$

denoting by Id the identity mapping on $\mathbb{R}^{3N \times 3N}$ and by $\mathcal{M}$ the blockdiagonal matrix

$$\mathcal{M} = \begin{pmatrix} \mathcal{M}_1 & & 0 \\ & \ddots & \\ 0 & & \mathcal{M}_N \end{pmatrix}$$

with $\mathcal{M}_{n,ij} = \varepsilon_{ijk} \hat{m}_{n,j}$, $n = 1,...,N$.

4.3 Magnetization dynamics

Further, the following vectors have been employed

$$\frac{\partial \tilde{\boldsymbol{m}}}{\partial t} = \frac{\partial}{\partial t}(\hat{m}_{x,1}, \hat{m}_{y,1}, \hat{m}_{z,1}, \hat{m}_{x,2}, \ldots)^T$$

and $\boldsymbol{H}_{\text{eff}} = (H_{\text{eff},x,1}, H_{\text{eff},y,1}, H_{\text{eff},z,1}, H_{\text{eff},x,2}, \ldots)^T$.

In this way, the solution of dynamic equations (3.15) applied on the N single-domain paticle system simplifies to an integration of equation (4.6) in respect to time. A similar approach was employed by D. Laroze et al., 2009, in order to investigate the dynamics of two interacting magnetic dipoles. Our calculations show identical results for this case. A typical magnetization evolution for the case of two dipolar coupled magnetic nanoparticles of diameter 20 nm with a centre distance of 25 nm can be found in Figure 4.5 if magnetic parameters of M_S = 1000 kA/m and α = 0.01 are assumed. The left plot shows the trajectory of the normalized magnetic moments. Since both curves converge to the same point, they align parallel along the symmetry axis of the system.

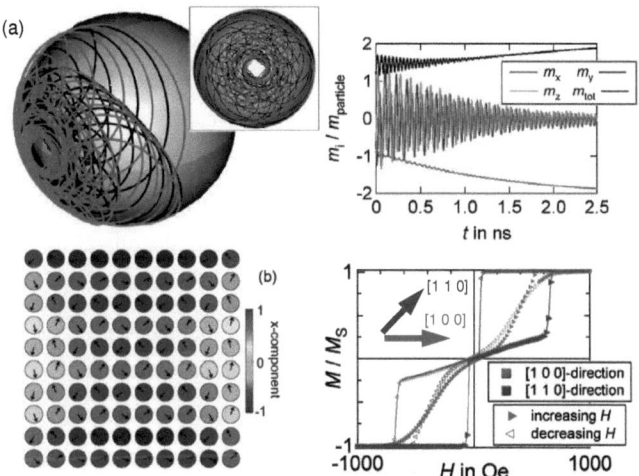

Figure 4.5: (a) Example for the dynamic behaviour of magnetic single domain nanoparticles. The left side shows the moment vector trajectories of two 20 nm particles with of a distance of 25 nm, a saturation magnetization M_S = 1000 kA/m and a damping coefficient α = 0.01. The evolution of single components can be found in the plot on the right side. (b) magnetic equilibrium and properties of a nanoparticles assembly. Depending on the direction of an external field, different $M(H)$-behaviour can be observed.

4. Interacting particles

4.4 Dipolar driven demagnetization processes

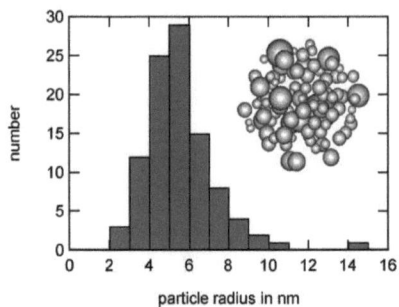

Figure 4.6: Particle size distribution of 100 particles according to the logarithmic normal distribution with expectation value of $\langle R \rangle = 6$ nm and $\sigma = 2$ nm. The inset shows the spatial particles distribution on a sphere of radius $R = 50$ nm.

In order to investigate the influence of the dipolar coupling on the demagnetization dynamics, a geometrical setup as shown in Figure 4.6 is chosen: small magnetic nanoparticles are equally randomly distributed across the volume of a three-dimensional sphere of radius R_S. The radii R of the magnetic nanoparticles follow a logarithmic normal size distribution

$$\rho(R) = \frac{1}{Rb\sqrt{2\pi}} \exp\left(-\frac{\ln R - a}{2b^2}\right) \tag{4.8}$$

with an expectation value $\langle R \rangle$ and a standard deviation σ given by

$$\langle R \rangle = \exp(a + \tfrac{1}{2}b^2)$$
and $\quad \sigma = (\exp(b^2) - 1) \cdot \exp(2a + \tfrac{1}{2}b^2),$

respectively. For each particle, we assume a uniaxial anisotropy with easy axis vector $\hat{e}_i$. The orientation of easy direction is chosen equally randomly along the surface of the two-dimensional unit sphere. For this particular case, the anisotropy functional is given by

$$f_{\text{ani}}(\hat{m}) = K(1 - \langle \hat{e}, \hat{m} \rangle^2) \quad \Rightarrow \quad \frac{\delta f_{\text{ani}}(\hat{m})}{\delta \hat{m}} = -2K \langle \hat{e}, \hat{m} \rangle \hat{m} \tag{4.9}$$

For the geometrical setup, we choose $N = 100$ nanoparticles. As an initial condition, the magnetization vectors are assumed to point in z-direction. This coincides with the experimental situation of particles in a strong external magnetic field which is switched off at $t = 0$. The external magnetic field acting on a particle is obtained by the summation of the stray field contributions of neighbouring particles according to the dipolar expression (4.2). We restrict the summation to all particles of a distance

4.4 Dipolar demagnetization

smaller than five times the average particle radius coinciding with the findings by Schaller et al., 2009b. We refer to the total magnetic moment of the multi-core particle at time t by $m(t)$ and to its components by m_x, m_y and m_z, respectively.

The trend in magnetic nanoparticle synthesis works to create particles with higher magnetization which ensures better particle handling in many applications e.g. magnetic separation or detection. Saturation values of around M_S = 1000 kA/m have been reported in 12 nm Co-particles (Ennen, 2008) which will be chosen as a reference here. The application of our findings to lower magnetizations (e.g. magnetite or maghemite, $M_S \approx 350$ kA/m), will be discussed below. Setting further $\alpha = 0.005$, $K = 0$, $\langle r \rangle = 6$ nm, $\sigma = 2$ nm, and $R = 50$ nm, the evolution of the total moment m is presented in Figure 4.7. The inset shows the typical behaviour of the normalized magnetization components of a single object inside the super structure.

As a matter of quantification, we introduce the typical decay duration τ via

$$m(\tau) = m(t = 0) \cdot \exp(-1) \tag{4.10}$$

In case of a single particle with a uniaxial anisotropy, the relaxation time τ is a function of the damping coefficient showing a minimum at $\alpha = 1$ (Figure 4.8, dashed line) (Russ and Bunde, 2006)

$$\tau \sim \frac{1+\alpha^2}{4\alpha} \tag{4.11}$$

However, due to its substructure a magnetic multi-core particle exhibits deviating dynamics. In Figure 4.8, the influence of different parameters on the relaxation time τ is presented. Each data point corresponds to an average of $\log \tau$ of 25 calculations with independently generated geometries. The error $\Delta \log \tau$ is calculated from standard deviation. Increasing the saturation magnetization M_S by a given factor leads to an equivalent increase of the logarithmic relaxation

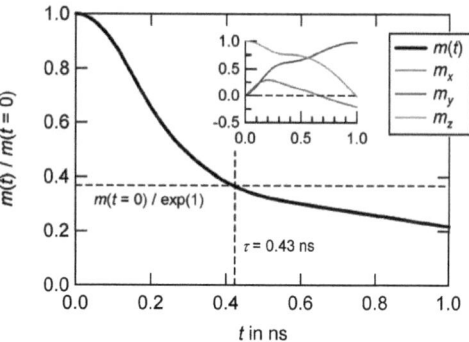

Figure 4.7: Dynamic relaxation of the total magnetic moment of a structure as shown in Figure 4.5 for the parameters M_S = 1000 kA/m, $\alpha = 0.005$, $\langle R \rangle = 6$ nm, $\sigma = 2$ nm and $R = 50$ nm, the inset shows the behaviour of a single superparamagnetic particle inside the structure.

4. Interacting particles

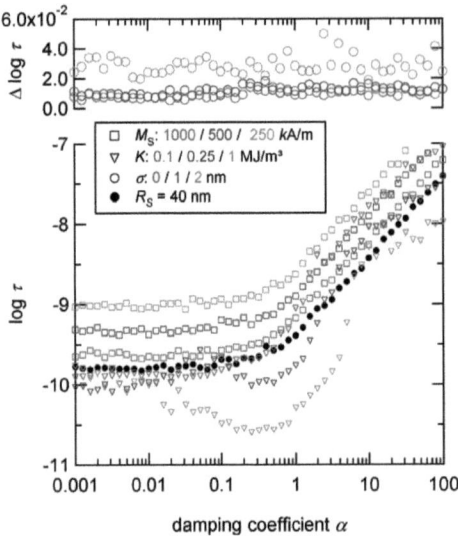

Figure 4.8: Dependence of the relaxation time on different geometry and material parameters. In all calculations $\langle R \rangle = 6$ nm is chosen. If the values for the remaining parameters are not explicitly given, they are $M_S = 1000$ kA/m, $K = 0$, $\sigma = 2$ nm and $R_S = 50$ nm.

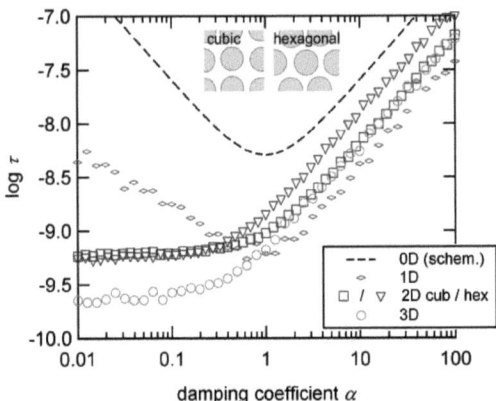

Figure 4.9: Relaxation of multi-core systems of different dimensions. In the case of high damping coefficients, $\alpha > 1$, the behaviour is similar, though relaxation is faster for a higher value α. For small damping coefficients the relaxation time τ depends on the dimension of the structure.

time $\log \tau$. A similar effect can be observed by varying the radius R_S of the multi-core particle; the relaxation time increases with increasing distance between nanoparticles. From this result, we can also extrapolate to the lower magnetization cases such as magnetite or maghemite, which scale by a factor given by the relation between the saturation magnetizations. The deviation of the times is directly related to the deviation of the particle size distribution. Choosing particles of a fixed radius $R = 6$ nm, we find a relaxation time almost independent of the geometry configuration. For higher σ the error $\Delta \log \tau$ increases.

A strong magnetic anisotropy decreases the relaxation time and a minimum of τ can be found. This minimum occurs at $\alpha = 1$ if the anisotropy becomes the dominating driving force within the system. The relaxation dynamic increasingly resembles the behaviour of a single particle (Figure 4.9, dashed line) though far smaller relaxation times can be observed for very high and very low damping coefficients for multi-core particles. This is due to the random orientation of the easy axis e of each nanoparticle. However, since such demagnetization processes are no longer dipolar driven, they are beyond the scope of our discussion. In the following we will therefore address systems with $K = 0$, only. In these cases, the curve shape remains unaffected, independent of the chosen parameters. For high damping constants, the dynamic behaviour of the three-dimensional system coincides with the relaxation of a single particle with uniaxial anisotropy (Fig. 4.9,

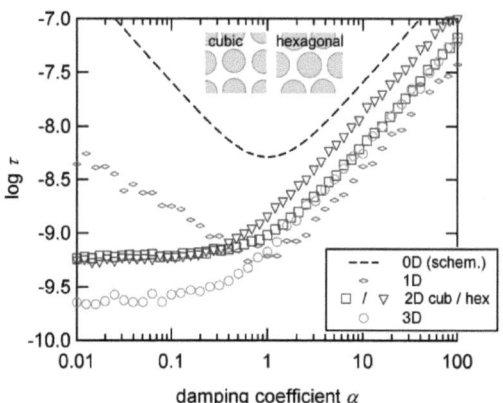

Figure 4.9: Relaxation of multi-core systems of different dimensions. In the case of high damping coefficients, $\alpha > 1$, the behaviour is similar, though relaxation is faster for a higher value α. For small damping coefficients the relaxation time τ depends on the dimension of the structure.

dashed line). The low damping cases strongly differ from this situation. The relaxation time τ is independent of the damping coefficient α.

To gain a better understanding of the governing dynamics causing an α-independent τ, we investigate similarly assembled systems on lower dimensions. Assuming again particles of $M_S = 1000$ kA/m, $K = 0$, $R = 6$ nm, we analyze the following particle patterns: (1) the three-dimensional spatial distribution as considered above, (2) a two-dimensional 10×10-cubic/hexagonal lattice of lattice constant 15 nm in the x-y-plane, (3) a one-dimensional particle array of 25 particles along the x-axis at a distance of 15 nm. In the cases of (2) and (3), we add a random spatial distortion of a mean value of 0.25 nm in random directions. The initial state of the magnetization is chosen to be in positive y-direction. In the case of (3), a switching of the magnetization direction from y- to x-direction is expected. Thus, the total moment will not decay and the definition of τ needs to be slightly modified. Instead of the decay length, we analyse the increasing duration of the m_x-component:

$$m_x(\tau) = m_y(t=0) \cdot (1 - \exp(-1)) \qquad (4.12)$$

This is in accordance with (4.12) in the sense that both definitions of τ coincide with the time when the magnetization has reached the limit of $t \to \infty$ up to a $1/e$-fraction.

4. Interacting particles

The results are shown in Figure 4.9: The one-dimensional particle chain shows a behaviour similar to that of a single particle with uniaxial anisotropy (Kronmüller, 2007). Qualitatively this can be readily understood; though no magnetocrystaline anisotropy is considered in our calculations, the assembly of particles introduces an effective uniaxial shape anisotropy which originates from dipolar coupling between the single-domain nanoparticles. The systems can be considered comparable as the time τ increases with decreasing damping in the low damping regime, differing from the two- and three-dimensional configuration. This coincides with the finding by D. Laroze et al., 2009, for dynamic two particle systems.

The two-dimensional particle lattices follow the one-dimensional behaviour for damping coefficients $\alpha > 1$, but differ for the low damping cases. Here no minimum can be found and τ goes to an α-independent value. Different lattice symmetries lead to the same stationary τ value, the curves, however, are shifted in respect to their α dependency. Here the hexagonal lattice reaches stationary relaxation time for smaller α than the cubic grid. The shift is due to a different number of direct neighbours in the geometry, which is 4 for the cubic and 6 for the hexagonal structure. This is also in accordance with the three-dimensional structure where higher numbers of neighbours (between 6 (cubic) – 12 (hexagonal)) can be achieved. A stationary τ behaviour is found for even smaller α.

According to our findings, the relaxation should occur along different paths depending on the system dimension. In the range of high damping ($\alpha > 1$), the system is overdamped; the phase trajectory in the k-space is independent of the spatial ordering investigated. The velocity of each magnetic moment vector goes to zero without strong oscillations. For small damping constants, the occupied k-space volume obtains a certain structure depending on the dimension of the system (Figure 4.10(a)). The dynamics of the one-dimensional array take place in the plane perpendicular to the chain direction. The k-volume of two-dimensional lattices is restricted to an ellipsoid containing the one-dimensional subplane. Only the three-dimensional spatial configuration shows a full spherical symmetry in their dynamics.

Common expectation suggests that an individual moment should become stationary after a time, determined by the parameter α, because the damping coefficient can be interpreted as measure for the energy leaving the system. This is actually the case as shown in Figure 4.10(b) for a 10×10-cubic lattice. The trajectories of each moment reveal that the equilibrium is dynamic. For each k-state, there is a mirrored state; thus the total magnetization cancels out over time until the microscopic dynamic vanishes.

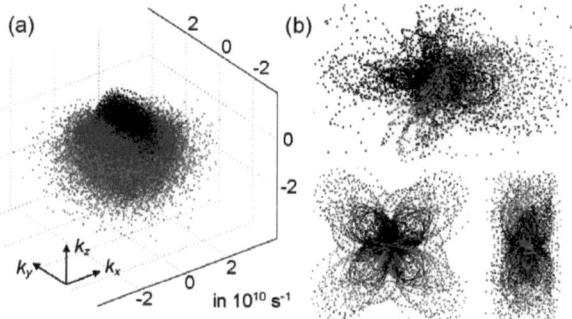

Figure 4.10: k-space volumes of different systems. (a) Cuts through the k-space trajectories for systems of different dimensions for $\alpha = 0.001$: 25 particle chain (black), 10×10-cubic lattice (blue), 100 particles of $\langle R \rangle = 6\,\text{nm}$, $M_S = 1000\,\text{kA/m}$, $K = 0$, $\sigma = 2$ nm and $R_S = 50$ nm (red). (b) Structure of k-volume for a 10×10-cubic lattice for $\alpha = 1$. For each state, there is a mirrored state; though on the microscopic level not stationary, the global structure has reached equilibrium. The red markers indicate the k-trajectory of a fixed particle in the array.

The situation in the three-dimensional case is similar, though there are even more possible configurations due to less spatial ordering. To conclude, we have investigated the dynamic demagnetization behaviour of magnetic multi-core particles in regards to different parameters. We have shown that the three-dimensional structures exhibit a dynamic behaviour that strongly deviates from single particle relaxations in the low damping regime. It may be remarked that these values of the damping coefficient are the only of physical relevance for common magnetic materials.

The observed behaviour could be related to an increase of the accessible volume in the k-space with increasing system dimension. One-dimensional particle chains qualitatively show a similar behaviour to single particles with uniaxial magnetocrystalline anisotropy. Their motion is confined to a two-dimensional subplane of the k-space, which leads to increasing relaxation time for decreasing $\alpha < 1$. This confinement is broken for systems of higher spatial dimension, resulting in relaxation time independent of the damping coefficient in the low damping regime. These findings are attributed to the fact that the macroscopic equilibrium for these dimensions is dynamic; the microscopic dynamics are still transient.

4. Interacting particles

4.5 Magnetic particles in adiabatically changed magnetic fields

We now return to the microfluidic devices discussed in chapter 2. One of the key assumptions when analyzing the separation device in section 2.2 was that we are dealing with sufficiently low particle concentrations. Considering superparamagnetic magnetic beads, this requirement is not important as long as particles are not brought into an external field. However, the situation changes if external fields come into play. Even if homogeneous fields are considered, the magnetic moments of the particles begin to align parallel to the field direction. For sufficiently high concentrations, such fields can influence nearby particles; a strong particle-particle interaction due to the dipolar stray field of each individual object arises. These interactions may also introduce hydrodynamic forces evoking a different behaviour than the analysis of free particles would predict (Mikkelsen et al., 2005a; Sawetzki et al., 2008). In such fields, magnetic particles begin aligning in one-dimensional chain arrays as shown in Figure 4.11. Since the dipolar attraction between neighbouring particles is very strong, chains do not break under fluid-induced shear stress. Instead, the chain will turn until an equilibrium configuration is reached, for strong magnetic fields the chain direction will coincide with the direction of the magnetic field.

If the size of such chains is sufficiently smaller than the geometrical size scale of the device, the chain will travel as a confined object through the microfluidic struc-

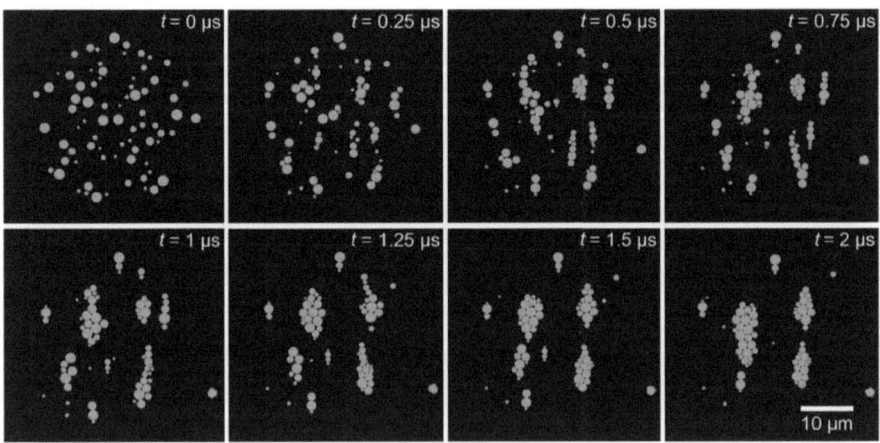

Figure 4.11: Agglomeration of particles of sizes between 0.5 and 1 μm in a homogeneous magnetic field parallel to the y-axis. The saturation magnetization of the particles is set to 1000 kA/m, their initial positions are chosen randomly on a two-dimensional sphere of radius 25 μm.

4.5 Particles in magnetic fields

ture. Except for some smaller corrections on the velocity and the angular momentum, the behaviour does not differ from the single particle case so long as no other (inhomogeneous) fields are considered. However, if chain length and geometrical size scale are comparable, different effects can be observed. In recent works of M.A.M. Gijs (Lacharme et al., 2008, 2009; Sivagnanam et al., 2009) chains were capture aligned perpendicular to the flow direction to realize sandwich immunoassays. However for continuous flow devices, such interactions have been regarded as negative side-effects which decrease the yield of devices. Contrary to these, we will employ such interactions for the manipulation of the particle flow without changing the fluid flow or applying inhomogeneous magnetic fields introduced by components on the microscale. Similar to the approaches discussed in the second chapter, the focus lies on keeping the device as simple as possible to make it applicable not only for laboratory system but also for many lab-on-a-chip devices.

The main component of the particle flow control device is schematically shown in Figure 4.12. A round reservoir leads to a rectangular channel. A homogeneous magnetic field $\boldsymbol{H}$ is applied along the whole microfluidic structure. The magnetic moment of diluted particles will align with the field direction and a particle j will feel a force due a particle i according to

$$\boldsymbol{F}_{mag}^{ji} = \mu_0 (\boldsymbol{m}_{part}^{j} \cdot \nabla) \boldsymbol{H}_{part}^{i} \qquad (4.13)$$

with $\boldsymbol{H}_{part}^{i}$ the dipolar particle stray field of particle i according to (4.2).

Thus, particles will start to agglomerate in chains oriented parallel to the external field. The average chain length depends on the concentration in the reservoir. A sufficiently long chain can only pass the junction area if the angle between chain and fluid flow direction is very small, otherwise it will be blocked at the junction. Smaller chains reaching the junction from the lower or the upper part of the geometry will feel a

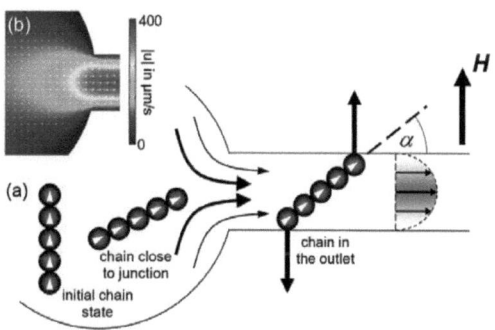

Figure 4.12: Schematic representation of the investigated device. A circular inlet reduces to a straight rectangular channel. The orientation of a travelling particle chain changes depending on the acting forces.

95

4. Interacting particles

torque due to the flow profile rotating their orientation parallel to the direction of the outlet channel as schematically shown in Figure 4.12(a). Fig. 4.12(b) shows a finite element calculation of the fluid profile in the junction area according to the Stokes equation (2.10). Magnetic and hydrodynamic torques always have opposite sign. Thus, if they are of similar size, smaller chains might pass the junction area and pass into the outlet. However due to a symmetric Poiseuille flow profile, the hydrodynamic torque vanishes along the outlet. The particle flow is blocked though the hydrodynamic flow remains. Critical conditions for the blocking of chains depend on the chain length L, the angle α between chain orientation and the external magnetic field H, and the magnetic moment of the particle chain.

The dependencies can readily be deduced: Since the rotation is blocked by the channel wall, the total torque needs to vanish. The turning of the particle chain is stopped by a force F

$$|F| = \frac{2}{3}\pi r^2 M_S \cdot \mu_0 |H|. \qquad (4.14)$$

The direction of the force depends on the angle α between the chain direction and the magnetic field. Denoting the channel width by h, the force direction can be calculated by $\cos\alpha = h/L$. The critical force necessary to block the chain depends also on properties of the channel wall. However, from (4.14) we can qualitatively conclude that for high enough field strengths $|H|$ and magnetizations M_S, it is always possible to fix the chain within the channel. Experimentally, it was shown that such blocking values can be reached without difficulties.

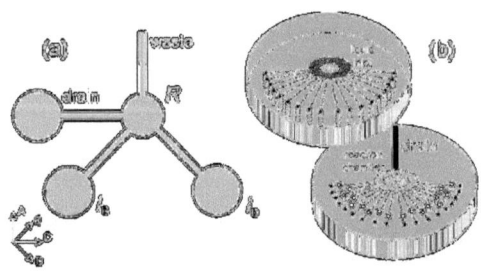

Figure 4.13: (a) Schematic of the investigated microfluidic device. It consists of two inlet reservoirs I_B and I_D, a reaction chamber R and a waste outlet. If the direction of the homogeneous magnetic field is chosen as B or D particles flow from the corresponding inlet into the chamber R. For directions according to A or C the waste or the drain is opened, respectively. (b) A possible improvement of the device: multiple channels are flooded by a single liquid reservoir; particles get into solution via particles reservoirs on a second bottom disc.

With this approach, different applications are possible. As an example we regard the structure shown in Figure 4.13(a) which serves as a particle diverter. Two

4.5 Particles in magnetic fields

inlet reservoirs I_B and I_D lead via straight channels with different orientations into a reaction chamber R. This chamber has two outlets: a drain reservoir for the collection of the reaction product and a waste that serves for initialization processes. If either orientation B or D is chosen, a particle flow from the corresponding inlet can be observed. This flow runs until the configuration is changed by opening the opposite valve; by washing the reaction chamber R; by applying the orientation A; by unlocking the waste or by transporting the particle mixture into the drain reservoir by applying a field in direction C. Basically, it is possible to extend this geometry to a higher number of inlets. Combining the structure in Figure 4.13(b) with a rotating magnetic field defines a cycle in which a specified number of particles always enter the reaction chamber. One turn of the field corresponds to two filling and emptying cycles of the reaction chamber.

Experiments on these systems have been carried out by F. Wittbracht. The channel geometry is realized using standard optical lithography and soft-lithography methods. In a first step, a negative of the designated fluidic geometry is produced by optical lithography of SU-8 3025 on a siliconoxide-terminated silicon wafer. Baking steps and exposure doses are chosen according to manufactures' instructions. This SU-8 structure serves as a mold mask in the second step of the sample preparation. Second, the polydimethylsiloxane

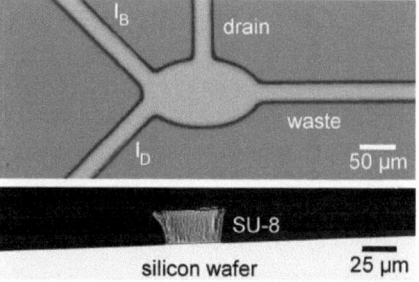

Figure 4.14: Optical microscopy image of the microfluidic device used in the experiments. The centred reaction chamber has a radius of 50 μm. All channels are 30 μm wide. The channel height of 25 μm is determined by a cross section image of the mold mask.

(PDMS) polymer kit is thoroughly mixed employing a 1:10 mass ratio of curing agent to silicone elastomer. After mixing the polymer solution, the mold mask is covered with the PDMS mixture. Afterwards the PDMS is cured at 80°C for 4.5 hours. After stripping the PDMS-layer off the substrate and trimming the channel structure, reservoirs are cut and the channels are cleaned in an ultrasonic bath. A siliconoxide-terminated silicon wafer serves as a channel bottom plate. To ensure proper sealing of the microfluidic geometry, a plasma oxidation of the PDMS structure and the silicon wafer is carried out, which in turn leads to the formation of an irreversible seal between the PDMS and siliconoxide surfaces (Jo et al., 2000). The resulting microflu-

4. Interacting particles

idic device is shown in Figure 4.14(a). Reservoirs have a radius of 500 µm, the channel dimension are given by a length of 2 mm, a width of 20 µm and a height of 25 µm. Channels are arranged in a rotational symmetry originating from the reaction chamber. Instead of a circular geometry, we chose an elliptic shaped chamber R to ensure better percolation. The chamber has semiaxes of 160 µm and 80 µm.

The experimental setup consists of a digital optical microscope (VHX-600, Keyence) with a built-in CCD-camera producing up to 28 frames per second. The sample is positioned on a pivotable sample holder which is surrounded by a pair of coils for the generation of a homogenous magnetic in-plane field. The field strength can be adjusted up to 490 Oe, while the orientation of the field direction can be adjusted by turning the sample stage. Since agglomeration along the channels can be expected (as per the discussions in the preliminary section), superparamagnetic particles need to be used in order to ensure the dissolving of clusters when the field direction is set to flow direction. To distinguish particles flowing from different inlet reservoirs into the reaction chamber, different magnetic objects are used: The bead-carrying reservoirs I_B and I_D are filled with solutions of Dynabeads MyOne^{TM} and Dynabeads M-280 (Fonnum et al., 2005) at concentrations of 10 mg/ml. Both bead species have narrow size distributions with standard deviations lower than 2%. Due to their different diameters of 1.05 µm and 2.8 µm, respectively, they can easily be kept apart by optical microscopy.

At the beginning of the experiment, the drain channel is filled with de-ionized water until the microfluidic device is completely filled. The bead reservoirs are subsequently filled with the corresponding bead solutions. During the filling process, the magnetic field of 490 Oe is aligned parallel to the drain channel. Due to dipolar interactions of the magnetic beads and the resulting chain formation aligned to the external field, no particle flow can be observed. By changing the relative orientation of the magnetic field and the microfluidic device, the bead flow can be manipulated. Figure 4.15(a-b) represents the case of field orientation D: Particles (M-280) begin to flow into the reaction chamber from the inlet I_D at a velocity of 100 µm/s. As long as the magnetic field vector points in this direction, no flow of MyOne^{TM} bead chains can be observed. Changing the relative orientation of microfluidic geometry and applied magnetic field (Figure 4.15(c)) stops the particle flow of M-280 particles. Aligning the field direction with the orientation B results in a flow of MyOne^{TM} bead chains from the inlet I_B into the reaction chamber (Figure 4.15(d-e)) while all other particles fluxes e.g. from I_D or into waste or drain are inhibited. The merging of incoming bead chains with chains deposited in the reaction chamber in preliminary procedure steps can be observed; this assembly is presented in Figure 4.15(f). After the formation of a bead chain consisting of M-280 and MyOne™ beads, the orientation of magnetic field and sample is changed again, leading to a parallel alignment of drain channel and magnetic field. During the rotation of the sample, the assembled bead chain breaks apart. This effect is due to the interaction of the bead chain with the reaction chamber wall. While one end of the chain is fixed at the contact point to the wall and can consequently no longer align in field direction, the opposite end follows the external field. This leads to high stresses along the chain centre which in the end cause the chain to break apart. The small fragment of the bead chain remains in the reaction chamber, whereas the large fragment is transported out of the reaction chamber as shown in Fig. 4.15(g-i), if the appropriate magnetic field direction is chosen. To prevent chain fragments from remaining in the reaction chamber, further optimization of the chamber shape is necessary.

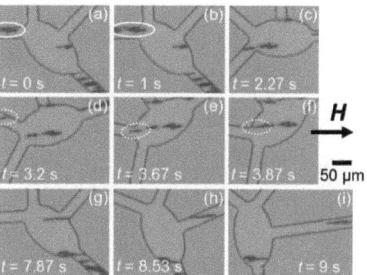

Figure 4.15: Microscopy images of the microfluidic device during operation. Particle flow can be controlled by the orientation between fluid flow and external field direction.

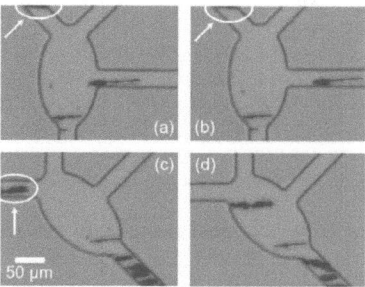

Figure 4.16: Microscopy images of chain blocking within the connecting channels. The marked chain is blocked in (a) and (b) but starts moving to the reaction chamber as soon as the field and flow orientation are parallel to each other

The gating of particle flow for certain magnetic field directions is shown in Figure 4.16. If the magnetic field is aligned parallel to the drain channel, a bead chain consisting of M-280 is kept from entering the reaction chamber, due to bead chain wall interactions as explained by (4.14). As displayed in Figure 4.16(a-b), the particle flow into the reaction chamber is inhibited, while the chamber is emptied into the drain. A chain of M-280 is blocked within the transport channel. Changing the orientation of the sample and the magnetic field, enables the flow of the previously detained M-280 bead chain into the reaction chamber as shown in Figure 4.16(c-d).

4. Interacting particles

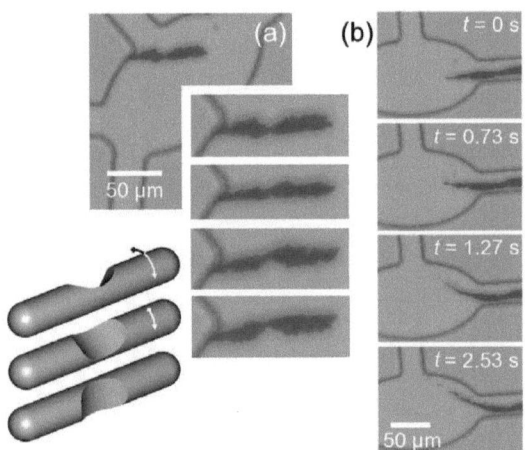

Figure 4.17: Additional effects enhancing the functionality of the device. (a) A particle chain reaches a blocked configuration close to one of the exits of the reaction chamber R, its stability is increased by self-ordering of particle positions within the fluid flow. (b) Switching off the external magnetic field, the particle chains dissolve within a time span of several seconds.

Other beneficial effects can be observed for the operation of the device. Due to rotational degrees of freedom, chains can rotate along their axis within the fluid flow to attain higher stability. This effect is shown in Figure 4.17(a), a particle chain reaches the channel wall and enhances stability by adjusting its spatial configuration to the fluid flow by rotation as shown schematically in Figure 4.17(a). Since superparamagnetic beads are employed, the switching off of the external magnetic field leads to the collapse of bead agglomerations and therefore enables the flow of individual particles. The deagglomeration is presented in Figure 4.17(b), the dissolving time for this process is given by only a few seconds. Thus, a bead mixture can be prepared in the reaction chamber and can afterwards be employed without additional preparations for further applications.

It needs to be pointed out again that the major advantage of this method is that only a homogeneous macroscopic magnetic field is necessary. In contrast to existing particle diverters (see e.g. Pekas et al., 2005) no electric components on the microscale are necessary which makes the device were easy to handle and in particular sufficient for the integration in existing lab-on-a-chip devices.

4.6 Conclusion and Outlook

In this chapter, we have investigated the influence of dipolar particle coupling with the help of a simplified version of the Landau-Lifshitz equation. For simplification, it was necessary that particles are homogeneously magnetized which made it possible to rewrite the original set of partial differential equations into a set of ordinary ones.

In section 4.4, we studied the dynamic demagnetization behaviour of magnetic multi-core particles in regards to different parameters. We have shown that the three-dimensional structures exhibit a dynamic behaviour strongly deviating from single particle relaxations in the low damping regime. It may be remarked, that these values of the damping coefficient are the only ones of physical relevance for common magnetic materials. The observed behaviour could be related to an increase of the accessible volume in the k-space with increasing system dimension. One-dimensional particle chains qualitatively show a similar behaviour to single particles with uniaxial magnetocrystalline anisotropy. Their motion is confined to a two-dimensional subplane of the k-space, which leads to an increasing relaxation time for decreasing α for $\alpha < 1$. This confinement is broken for systems of higher spatial dimension, resulting in relaxation times independent of the damping coefficient in the low damping regime. These findings are attributed to the fact that the macroscopic equilibrium for these dimensions is dynamic; the microscopic dynamics are still transient. These results have been accepted for publication in *Dynamic simulations of the dipolar driven demagnitization process of magnetic multi-core nanoparticles*, J. Magn. Magn. Mat., in press.

In section 4.5, it was shown that dipole-dipole interaction between magnetic beads can be employed to manipulate the particle flow in continuous flow devices by applying a homogenous magnetic field only. Particles can be restricted to areas without changing the state of motion of the liquid and without the integration of electromagnetic components on the microscale which should be a major advantage of this setup in comparison to existing diverters. Additionally, captured particle chains can be released without any delay by aligning field and fluid flow direction. The geometry investigated can easily be extended to a particle diverter of numerous inlet reservoirs for mixing and reaction applications and due to its simplicity, it can be easily implemented into existing devices. To our knowledge the predictions and realizations presented here, show the first direct employment of dipolar particle-particle interactions

4. Interacting particles

in continuous flow devices. The results entitled *Particle flow control by induced dipolar particle interactions* are currently under consideration at Microfluid. Nanofluid.

For the discussion in section 4.5, we only considered the case of magnetic fields that change sufficiently slowly so the chain axis can always follow the field orientation. If very high frequencies are considered though, very strong hydrodynamic forces act on the chain segments. In particular, this might lead to the breaking of the assembly followed by a repulsive force between the resulting two segments which is due to dipolar interaction. Mikkelsen et al., 2005b, already reported that in situations where dissolved particles interact with each other via dipolar coupling also very high hydrodynamic forces can be observed. In the framework of this thesis, different aspects of such phenomena have been investigated. However, the analysis of these systems is still work in progress; therefore, we will only give a short outlook.

4.6.1 Outlook: Magnetic particles in high frequency fields

In this model, it is not sufficient to discuss the linear Stokes equation for the description of the liquid properties but instead the full Navier-Stokes equation needs to be considered. Though a model on the microscale is investigated, very high local Reynolds numbers can be reached. It is therefore necessary to have an appropriate mesh resolution near to the particles themselves. Similar as to the works of V. Thümmler and W.J. Beyn et al., 2004; 2008; 2009, the mesh is transported in respect to the particle motion. The implementation is done in an ALE-framework (compare section 1.3). For the definition of the ALE-mappings $\mathcal{A}_t : \Omega_0 \rightarrow \Omega_t \cong \Omega_0$, we introduce a second domain triangulation $\mathcal{T}_2$: the inner nodes $\mathcal{N}_{int}$ are indicated by the positions, additional auxiliary nodes need to be specified along the domain boundary. From the set of nodes, we obtain the triangulation using *Delaunay algorithm*. The result is presented in Figure 4.18(a).

The ALE-mappings can now be defined by employing standard finite element techniques (compare section 1.2.4): Denoting the nodes of a triangle T by r_1, r_2, r_3 with $r_i = (x_i, y_i)^T$, T can be parameterized by parameters s_1 and s_2 via mapping the triangle T onto the two-dimensional simplex S_2. Using the affine mapping

$$\Phi_T(s_1, s_2) = r_1 + s_1(r_2 - r_1) + s_2(r_3 - r_1) \tag{4.15}$$

it can be shown that

$$s_1 = \frac{(x-x_1)(y_3-y_1)-(x_3-x_1)(y-y_1)}{(x_3-x_1)(y_3-y_1)-(x_2-x_1)(y_3-y_1)} \quad (4.16a)$$

$$s_2 = \frac{(x-x_1)(y_2-y_1)-(x_2-x_1)(y-y_1)}{(x_3-x_1)(y_2-y_1)-(x_2-x_1)(y_3-y_1)} \quad (4.16b)$$

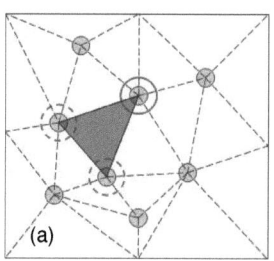

Assuming the mesh velocity is (a) equal to the particle velocity on the particle domains and (b) changes linearly in between the particles, suitable functions to model the mesh velocity are given by the linear hat functions (compare Figure 4.18(b))

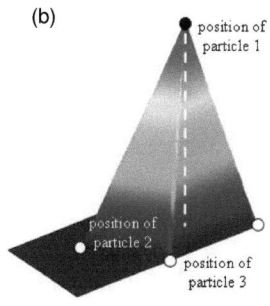

$$\Lambda_T(x,y) = 1-(s_1+s_2) \quad (4.17)$$
$$= \frac{(y_3-y_2)(x_3-x)-(x_3-x_2)(y_3-y)}{(x_2-x_1)(y_3-y_1)-(x_3-x_1)(y_2-y_1)}.$$

However, this choice leads to difficulties concerning numerical stability resulting from a rapidly decreasing element quality of the original FEM-mesh. To overcome these problems, slightly modified functions are applied:

Figure 4.18: Construction of the ALE-mapping via a second domain triangulation $\mathcal{T}_2$: (a) triangulation $\mathcal{T}_2$ with particle positions as inner nodes, (b) resulting linear hat functions.

$$\tilde{\Lambda}_T(x,y,\theta_1,\theta_2) = \frac{\Lambda(x,y)-\theta_1}{1-\theta_2-\theta_1}\Theta(\Lambda(x,y)-\theta_1)\cdot\Theta(1-\theta_2-\Lambda(x,y))$$
$$+\Theta(\Lambda(x,y)-(1-\theta_2)) \quad (4.18)$$

with $\theta_1, \theta_2 \in [0,0.5)$ two numerical parameters which lead to an acceleration of surrounding mesh elements and increase the mesh quality. A thorough analysis estimating the choice of θ_1 and θ_2 has not yet been accomplished, however, we find indications for a coupling to the element *growth rate* close to the particle positions. Denoting the ALE- or reference coordinates of a particle at spatial coordinate r_i by ξ_i, the displacement Δr of an initial coordinate r may be written by

$$\Delta r = \sum_i (r_i - \xi_i) \cdot \tilde{\Lambda}_T(x,y,\theta_1,\theta_2) \quad (4.19)$$

4. Interacting particles

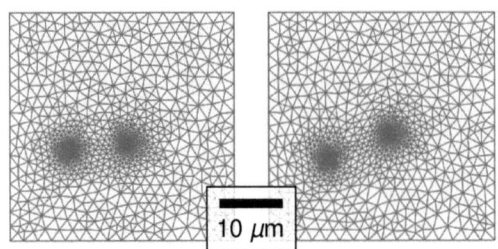

Figure 4.19: Moving mesh for two interacting particles. The ALE-approach maintains a high mesh resolution close to the particle positions. The plots show the initial configuration (left) and mesh displacement after a certain time (right).

with $\mathcal{A}_t(\xi) = \xi + \Delta r$ the ALE-mapping. The physical equations can be recast into the reference system according to (1.21). It should be pointed out, that in this case the ALE-formalism is not employed to deal with a moving domain as introduced in section 1.3. Instead it locally ensures a sufficient mesh resolution which is necessary for a proper discretization of the Navier-Stokes equation and the time-dependent dynamics. An example for two interacting objects is shown in Figure 4.19, the positions of the particles coincide with the areas of high resolution. The ALE-approach maintains the mesh resolution close to the particle positions.

Just as the application of Stokes equation is not correct here, we may no longer neglect inertia effects. Therefore, treating each particle as a point mass rather than applying (1.14) right away, the motions of the particles need to be calculated from Newton's second law

$$m_i \frac{dv_i}{dt} = F_{i,\text{fluid}} + F_{i,\text{mag}} + F_{i,\text{pen}} \tag{4.20}$$

with
- m_i mass of i-th particle
- $F_{i,\text{fluid}}$ fluidic forces acting on the i-th particle
- $F_{i,\text{mag}}$ magnetic forces acting on the i-th particle
- $F_{i,\text{pen}}$ penalty forces acting on the i-th particle preventing particle from overlapping

As locally high Reynolds numbers can be found, fluidic forces are modeled by the Khan-Richardson force (2.15), instead of applying Stokes drag law which only holds in the creeping flow regime. The magnetic force is given by $(m_i \nabla) H$ with m_i the particle moment and H the magnetic field at the particle position. Finally, for the penalty contribution, we choose a force originating from a *Lennard-Jones-potential*:

4.6 Outlook

$$F_{pen} = -\nabla \varphi_{LJ} \quad \text{with} \quad \varphi_{LJ}(r) = 4\varepsilon\left(\left(\frac{\sigma}{r}\right)^{12} - \left(\frac{\sigma}{r}\right)^{6}\right) \quad (4.21)$$

with ε the depth of the potential well and σ the distance at which the interparticle potential is zero.

A moving particle exerts a force on the liquid. Considering particles as point masses, this force is given by

$$f(r,t) = \sum_i f_i(t) \cdot \delta(r_i(t) - r)$$

Unfortunately, this expression introduces some technical difficulties. The Kronecker δ-distribution is no longer an L^2-function, the convergence of such numerical schemes therefore becomes problematic. For the implementation, we introduce instead additional geometry nodes, where we employ weak point terms.

First, calculations were carried out under the assumption of a particle mass density of 2500 kg/m³ and a saturation magnetization of 1000 kA/m. The carrier liquid is supposed to be water at room temperature, therefore, we set $\eta = 1.002 \cdot 10^{-3}$ Pa s and $\rho = 998.2$ kg/m³. Starting with only two particles of identical radius R, we observe the following frequency-dependent distance behaviour

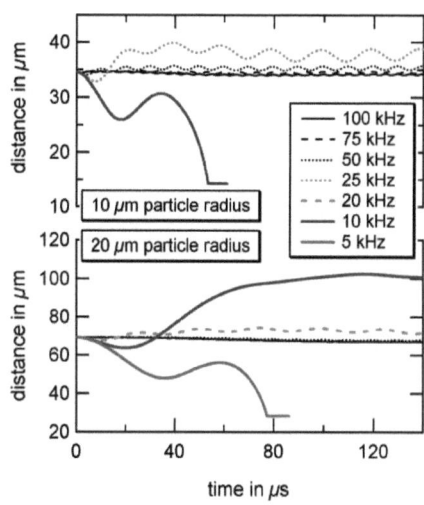

Figure 4.20: Behaviour of the particle distance in respect to the field frequency for the case of a particle of radius (a) 10 µm and (b) 20 µm.

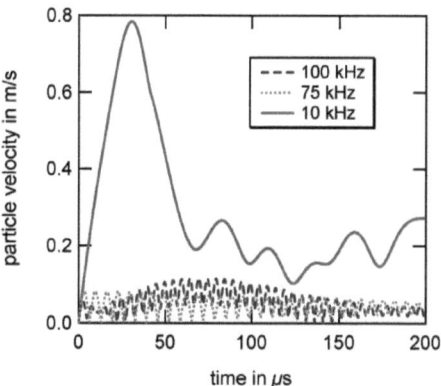

Figure 4.21: Particle velocities for particles of radius R = 10 µm for different field frequencies.

as presented in Figure 4.20: at low frequencies, the distance of the particles strongly decreases and remains constant (this can also be found experimentally). At very high frequencies, particles oscillate against each other. However due to a very rapidly

4. Interacting particles

changing magnetic field, the particles wander close to their initial position; the average distance in respect to time that a particle moves is close to zero. From Figure 4.20, we learn that there is clearly a critical frequency area, where particles are pushed apart (~ 25 kHz for particles of a radius 10 μm and ~ 10 kHz for particles of a radius 20 μm). This area changes if particles of different size are investigated, as can be seen in Figure 4.20(b). Similar results are obtained if different values for the magnetization M_S are discussed.

Regarding the motion of the particles, it is interesting to notice that the particles on the microscale can actually reach macroscopic velocities (Figure 4.21). Thus, it is no longer clear whether magnetic or hydrodynamic interaction is the main force contributing. For the comparison, we choose a system of two interacting magnetic markers of $R = 1$ μm and the remaining parameters as given above. If a frequency of 50 kHz is applied, such particles induce a flow profile as shown in Figure 4.22(a). A third probe particle of variable radius feels fluidic and magnetic forces. The dominant contribution depends on the particle position. Figure 4.22(b) shows an influence plot. Electromagnetic interactions dominate close to the particles, whereas in the outside ares, hydrodynamic forces play a major role. The bright regions in between correspond to intermediate regimes. Similar to the findings of C. Mikkelson et al., 2005b, we find a dominating magnetic contribution at short range.

Unfortunately, a thorough analysis of these systems is not yet finished at the moment. However, the simulation examples at this point already indicate that the modeling of particles immersed in fluid flows as "free particles" might lead to wrong re-

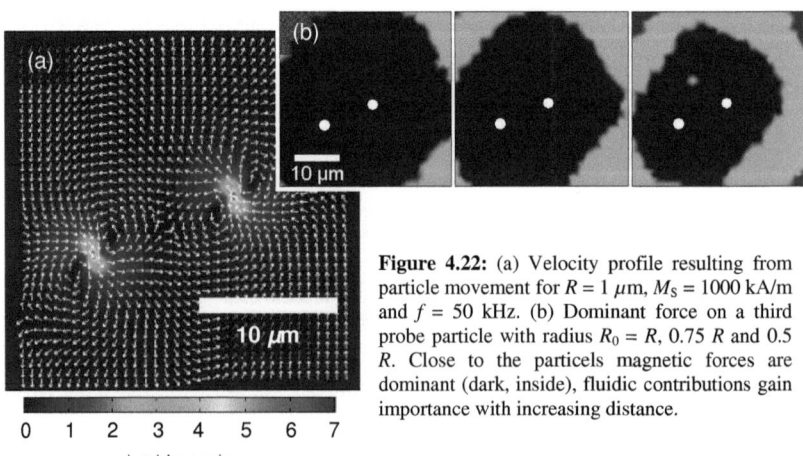

Figure 4.22: (a) Velocity profile resulting from particle movement for $R = 1$ μm, $M_S = 1000$ kA/m and $f = 50$ kHz. (b) Dominant force on a third probe particle with radius $R_0 = R$, 0.75 R and 0.5 R. Close to the particels magnetic forces are dominant (dark, inside), fluidic contributions gain importance with increasing distance.

sults if high concentrations in strong external fields are considered. The importance of different force contributions will be addressed in future works. A point of high interest in this regard will be the question whether non-linear effects can be observed. Such phenomena would have promising applications in the field of mixing in the laminar flow regime.

4. Interacting particles

Chapter 5

Detection of magnetic particles

After the discussion of the magnetic properties of magnetic beads and nanoparticles and their behaviour in continuous flow devices, we return to the originally stated problem: the detection of magnetic markers. Magnetic particles influence soft magnetic material nearby due to their magnetic stray field. This enables their detection by magnetoresistive sensors and thus an indirect visualization of every biological or chemical component bound to the marker (Wang and Li, 2008; Brückl et al., 2005). This strategy has been pursued during the last decade in applications involving biological recognition (Graham et al., 2003).

Different measuring tasks may be formulated. The simplest question to answer is if a magnetic bead or particle is close to the sensor. However, since the particle stray field is inhomogeneous, additional information on the position of the particle might be possible, i.e. instead of only asking for a yes/no-answer, we want to obtain (spatially highly resolved) position information. In regards to applications, it might also be interesting to know how many particles are within detection range. Since magnetic fields follow the principle of superposition, (i.e. the magnetic field of two magnetic particles is given by the sum of the individual particle stray fields) the effect of a single particle should be smaller than the evoked sensor response of several magnetic agents. Depending on the detection task, different requirements need to be met by the sensor setup. Number sensitive particle detection demands a sensors response independent of the particle positions. This commonly implies sensor dimensions far larger than the particle sizes (Liu et al., 2006). We will still show that the total signal sums

5. Particle detection

up linearly from the contribution of each particle as long as particle-particle interaction can be omitted. Thus, the magnitude of the measured signal refers uniquely to the number of particles. If instead a high space resolution is required, strong response changes in respect to the particle position need to be achieved which entails the employment of sensors of sizes equal to or smaller than the particle dimensions.

Both applications are limited by the influence of device noise, which introduces a threshold of the magnetic field strength that can still be detected. Since the magnetic stray field decays rapidly with the distance between particle and sensor, such thresholds introduce strong restrictions on the measurable signal (Kogan, 1998; Hedwig, 2009). As recently shown by J. Loureiro et al., 2009, dynamic measurements of particles in continuous flows are possible if sensor layouts and target particles of sufficient sizes and magnetic moments are chosen. However, this gives no guideline for the sensor design if particles on the nanometer size scale are considered.

The sensors discussed in this chapter can all be found on the size scale of < 1 μm. At these dimensions, the stray field coupling between neighbouring magnetic films is of major importance. In section 5.1, the model used for the simulation is discussed. We employ a two-dimensional approach for the calculation of the layer dynamics while the magnetic stray field is simulated in a three-dimensional frame. We will also briefly explain the implementation of our approach by the COMSOL plug-in PADIMA which was developed in the framework of this thesis (for a brief introduction refer to appendix A.4). In section 5.2, we test our approach investigating the behaviour of a vortex free layer state under the influence of an external magnetic field and compare the results with an analytic model. The description of single particle detection may be found in section 5.3. After establishing our model by comparison to the experimental data, we estimate the space resolution of such a magnetoresistive detector. In particular, we introduce a method how to construct sensor arrays that have detection precision below a given threshold with only a small amount of sensors employed. To overcome difficulties with the particle height due to a rapidly decaying stray field which is one of the main problems for the in flow detection (Loureiro et al., 2009), we introduce sensor geometries and show that it is possible to increase the detection range via highly sensitive areas. The chapter ends with analysis of the multi particle case, where we combine the model of dipolar-coupled homogeneously magnetized spheres to the sensor model. We will see that the particle signal can be decomposed into a linear contribution originating from free particles and a hysteresis one from coupled particles.

The main results of this chapter have been published in various scientific journals. The analysis of the experimental observations made by C. Albon in the framework of her PhD thesis (Albon, 2009) led to two publications. The interpretation of the measured signal for the case of a single MyOne™, presented in section 5.3.1, can be found in *Tunneling magnetoresistance sensors for high resolutive particle detection*, Applied Physics Letters **95** (2), 023101 (2009). The analysis of the influence of particle-particle coupling as discussed in section 5.4 have been published in *Number sensitive detection and direct imaging of dipolar coupled magnetic nanoparticles by tunnel magnetoresistance sensors*, Applied Physics Letters **95** (16), 163106 (2009). The estimation of the spatial resolution limits and the entailed design guidelines employing elliptically shaped sensors was released in: *On the resolution limits of magnetic sensors for particle detection*, New Journal of Physics **11**, 113027 (2009).

The results presented in section 5.3.3 have been submitted to Applied Physics Letters.: *Toward the magnetoresistive detection of single nanoparticle: new strategies for particle detection by adjustment of sensor shape* and are currently under consideration.

5.1 Weak formulation and thin film approach

For the implementation of the dynamic equations of micromagnetism, it is necessary to bring the original equation into the weak formulation. Starting from

$$\frac{\partial \hat{m}}{\partial t} = -\gamma \hat{m} \times H_{eff} - \alpha \hat{m} \times \frac{\partial \hat{m}}{\partial t}$$

with $\quad H_{eff} = -\frac{2A}{\mu_0 M_S}\Delta\hat{m} + \frac{2K_1}{\mu_0 M_S}\langle\hat{m},\hat{e}\rangle\hat{e} + H_{demag} + H_{ext}$,

we multiply with an arbitrary vector test function ψ and integrate over the whole magnetic domain Ω_{mag}. It needs to be pointed out that the maximum order of weak spatial derivatives allowed is ≤ 1 since the discretization is done in a Galerkin framework with $H^1(\Omega_{mag})$ as the solution space for weak solutions. Therefore, the first summand appearing in the effective field H_{eff} needs to be rewritten according to

$$\int_{\Omega_{mag}} \langle\psi,\hat{m}\times\Delta\hat{m}\rangle\, dr = \sum_{i,o} \int_{\Omega_{mag}} \psi_i \varepsilon_{ijk} \hat{m}_j \partial_o^2 \hat{m}_k\, dr$$

5. Particle detection

$$= \sum_{i,o} \int_{\Omega_{mag}} \left(\partial_o(\psi_i \varepsilon_{ijk} \hat{m}_j \partial_o \hat{m}_k) - \partial_o \psi_i \cdot \varepsilon_{ijk} \hat{m}_j \partial_o \hat{m}_k - \psi_i \varepsilon_{ijk} \partial_o \hat{m}_j \partial_o \hat{m}_k \right) dr$$

$$= \sum_{i,o} \int_{\Omega_{mag}} \left(\partial_o(\psi_i \varepsilon_{ijk} \hat{m}_j \partial_o \hat{m}_k) - \partial_o \psi_i \cdot \varepsilon_{ijk} \hat{m}_j \partial_o \hat{m}_k \right) dr \tag{5.1}$$

The last equality holds since the third summand is a product of the antisymmetric tensor ε_{ijk} and the symmetric one $\partial_o \hat{m}_j \partial_o \hat{m}_k$, therefore its overall sum vanishes. Integrating by parts and assuming a homogeneous van Neumann condition, (5.1) simplifies to

$$\int_{\Omega_{mag}} \langle \psi, \hat{m} \times \Delta \hat{m} \rangle dr = \int_{\partial \Omega_{mag}} \left\langle \psi, \hat{m} \times \frac{\partial \hat{m}}{\partial \hat{n}} \right\rangle dr - \sum_{x,y,z} \int_{\Omega_{mag}} \left\langle \left(\hat{m} \times \frac{\partial \hat{m}}{\partial x_i} \right), \frac{\partial \psi}{\partial x_i} \right\rangle dr$$

$$= -\sum_{x,y,z} \int_{\Omega_{mag}} \left\langle \left(\hat{m} \times \frac{\partial \hat{m}}{\partial x_i} \right), \frac{\partial \psi}{\partial x_i} \right\rangle dr \tag{5.2}$$

Thus, the weak form of (3.18) may be written as

find $\hat{m} : \Omega_{mag} \to S^3$, such that

$$\int_{\Omega_{mag}} \left\langle \psi, \frac{\partial \hat{m}}{\partial t} - \alpha \hat{m} \times \frac{\partial \hat{m}}{\partial t} \right\rangle dr = -\frac{2A}{\mu_0 M_S} \sum_{x,y,z} \int_{\Omega_{mag}} \left\langle \left(\hat{m} \times \frac{\partial \hat{m}}{\partial x_i} \right), \frac{\partial \psi}{\partial x_i} \right\rangle dr$$

$$+ \frac{2K_1}{\mu_0 M_S} \int_{\Omega_{mag}} \langle \psi, \hat{m} \times \hat{e} \rangle \langle \hat{m}, \hat{e} \rangle dr$$

$$+ \int_{\Omega_{mag}} \langle \psi, \hat{m} \times (H_{demag} + H_{ext}) \rangle dr \qquad \text{on } \Omega \quad (5.3)$$

$$\frac{\partial \hat{m}}{\partial \hat{n}} = 0 \qquad \text{on } \partial \Omega$$

Modelling systems with high aspect ratios in a finite element framework leads to complications: to maintain numerical stability and thus ensure convergence of the numerical scheme, triangles / tetrahedrons of high *element quality* need to be employed for the domain discretization. The element quality is a measure of how close the subdomains are to an equilateral triangle / tetrahedron. In the case of thin magnetic layers, this requires that the lateral domain decomposition is on the same scale

5.1 Weak formulation

as the perpendicular one which is bounded by the film thickness itself. Thus, a full three-dimensional treatment leads to a very high number of elements and, consequently, also to a large amount of degrees of freedom.

Therefore, in this work a different approach is chosen: due to magnetic exchange energy the size scale on which a change of magnetization components can be observed exceeds the thickness of the film by several magnitudes. Considering a magnetic layer in x-y-plane, we can approximate the distribution by $\hat{m}(x,y,z) = \hat{m}(x,y)$; the magnetic thin film can be modelled by a two-dimensional geometry. The sensors discussed in this work all have an area of less than 1 µm². For this type of setup, the coupling of magnetic films via their stray field is of major importance (Meyners, 2006). In order to ensure high exactness of the field approximation, a three dimensional approach is chosen. For practical calculations, we work in two frames: a three-dimensional one for the calculation of the magnetic potential ϕ_{mag} and a two-dimensional system for magnetic properties of the layers. It needs to be remarked that the calculation of the magnetic potential needs to be restricted to a finite domain. The size of the domain has to be adjusted so that the domain boundary does not influence the field along the magnetic domain. Due to a strong coupling between adjacent layers, a surrounding sphere of radius given by twice the radius of the circumscribing sphere of the sensor geometry is found to be sufficient.

Since for the evaluations of ϕ_{mag} in the three-dimensional system the evaluation of $\hat{m}$ is necessary and vice versa, we also need to introduce projection mappings connecting both frames with each other. Denoting the free magnetic layers in three dimensions by $\Omega^3_{mag,i}$ and the two-dimensional domain by Ω^2_{mag}, these projection mappings are given via

$$\Phi_i : \Omega^3_{mag,i} \to \Omega^2_{mag} \qquad \Phi_i^{-1} : \Omega^2_{mag} \to \Omega^3_{mag,i} \qquad (5.4)$$
$$(x,y,z) \mapsto (x,y) \qquad\qquad (x,y) \mapsto (x,y,z)$$

for every z such that (x,y,z) can be found in the i-th layer. Thus, the magnetization $\hat{m}$ in the two-dimensional frame is projected onto the three-dimensional layers via $\hat{m} \circ \Phi_i^{-1}$ whereas the potential in the two-dimensional coordinate system is given by $\phi_{mag} \circ \Phi_i$. A schematic representation of the modelling settings is shown in Figure 5.1: a magnetic multilayer system with a magnetization distribution initially aligned with the positive x-axis goes over to its equilibrium configuration. Due to the chosen dimensions (side length of 100 nm), the dynamics are dominated by the stray field

5. Particle detection

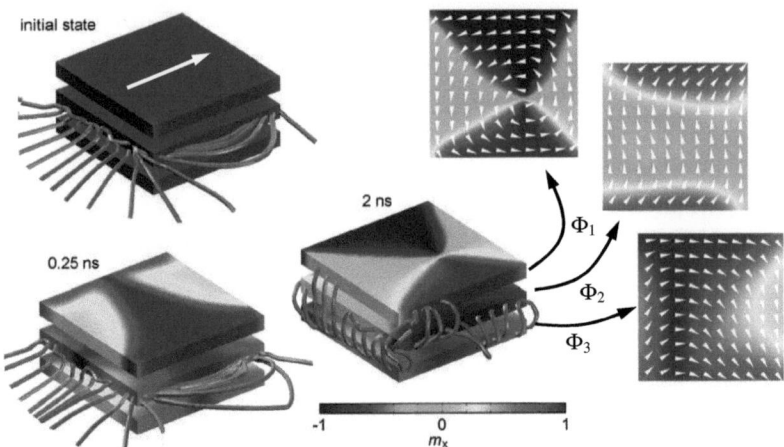

Figure 5.1: Schematic representation of the simulation approach. For the calculation of the magnetic stray field, a three dimensional approach is chosen. This frame is connected via projection mappings Φ_l on each layer with a two-dimensional geometry employed for the calculation of the magnetization dynamics. The example shows the relaxation of an initially homogeneously magnetized trilayer system. Due to the choice of the material parameters, different equilibrium states are reached. From top to bottom, we find vortex-, S- and C-state which are common solutions for such systems [HKro00].

coupling of layers. The equilibrium states of the layers are well-known solutions for micromagnetic systems. From top to bottom we find a vortex, an S- and a C-state (Kronmüller and Hertel, 2000).

5.1.1 Tunnelling magnetoresistance sensors

The sensors investigated in the following consist of ferromagnetic, conducting layers which are separated by an insulating tunnelling barrier. Due to tunnelling processes of electrons across the barrier, a tunnelling current can be measured that was first observed in 1975 by M. Julière in Fe/GeO/Co-junctions (Julière, 1975). The resistance along the device changes if an external magnetic field is applied, an effect that is called ***Tunnelling MagnetoResistance***(TMR)-effect. In particular, a low resistance R_p is found for parallel orientation of the magnetizations of the laye while a high resistance R_{ap} is obtained if layer magnetizations are oriented antiparallel. The TMR-ratio is defined by

$$\text{TMR} = \frac{R_{ap} - R_p}{R_p}. \tag{5.5a}$$

Julière related the tunnelling current to the so-called *spin polarization P* which is a measure for the spin relation at the Fermi level (compare Figure 5.2).

We denote the spin polarizations of top and bottom electrode by P_{top} and P_{bottom}, respectively. TMR-value and polarizations can be connected by applying a simple two current model in which a current running through the device is decomposed into electrons with different spins. The tunnelling probability of a spin with a given spin direction depends on the spin density at the Fermi-level in both electrons. The modelling of different cases is shown in Figure 5.2. In particular, by employing the spin polarization (5.5) may be re-written as

$$\text{TMR} = \frac{2 P_{\text{top}} P_{\text{bottom}}}{1 - P_{\text{top}} P_{\text{bottom}}} \quad (5.5b)$$

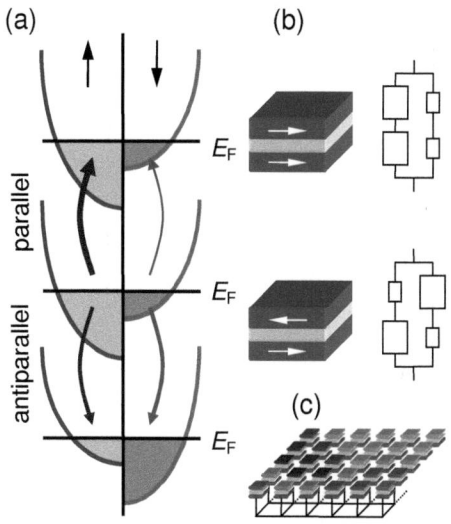

Figure 5.2: (a) Tunnelling processes depending on the spin polarization at the Fermi level E_F, (b) resistance approximation be two current model and (c) approximation of the continuous magnetization layer by a parallel connected array of homogeneous segments.

Considering a system of two ferromagnetic layers with layer magnetization $\hat{m}_1, \hat{m}_2$, respectively, the TMR-ratio needs to be connected to the effective values. We therefore think of the magnetic layers as an array of homogeneously magnetized layer stacks assembled in a parallel circuit. In this situation, the behaviour of each element is given by (5.5b) and depends on the relative orientation of the magnetization orientations. Employing the parallel connection, the equivalent resistance may be calculated and the TMR-value based on the effective variables is obtained. The TMR-value depends on the relative orientation of the magnetization vectors at every point in the film plane and is given by (Schepper et al., 2004)

$$\text{TMR} = p_m \frac{1 - \langle \alpha \rangle}{1 + p_m \langle \alpha \rangle} \quad (5.6)$$

5. Particle detection

with $p_m = \dfrac{\text{TMR}_{max}}{2 + \text{TMR}_{max}}$ and $\langle \alpha \rangle = \dfrac{1}{\| A_{mag} \|} \int\limits_{A_{mag}} \cos(\sphericalangle(\hat{m}_1, \hat{m}_2)) dr$

with TMR_{max} the maximum TMR-value obtained from experimental comparison and A_{mag} the interface area of the magnetic films.

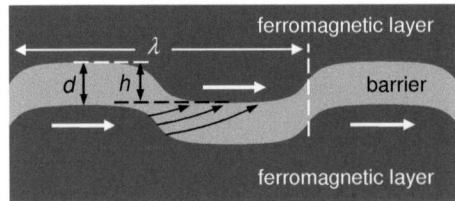

Figure 5.3: Layer coupling via correlated surface roughness. Néel coupling favours a parallel alignment of the magnetization in adjacent layers.

When dealing with very thin films, the stray field interaction between neighbouring magnetic layers is not the only coupling that needs to be considered. Due to the very thin non-magnetic spacer layer, roughness and defects within the surfaces are correlated as shown in Figure 5.3. A bump in the lower electrode induces a feature in the upper electrode and vice versa. A common approach to take these perturbations into account is to assume a sinusoidal roughness structure of wave length λ and amplitude h. According to Néel this leads to an additional surface energy $J_{\text{Néel}}$ (Néel, 1962) given by

$$J_{\text{Néel}} = \dfrac{\mu_0 \pi^2 h^2}{\sqrt{2}\lambda} M_S^2 \exp\left(-\dfrac{2\pi\sqrt{2}d}{\lambda}\right) \langle \hat{m}_1, \hat{m}_1 \rangle \qquad (5.7)$$

According to (3.16), this term acts along the boundary of the magnetic domains. We incorporate it into our model, by adding the additional term $C \cdot \delta J_{\text{Néel}} / \delta \hat{m}_i$ where the constant C transforms the surface energy into the corresponding volume term. This approach is reasonable as no change of the magnetization vector perpendicular to the film plane is expected.

5.1.2 COMSOL implementation: PADIMA

In many cases, we are only interested in the equilibrium state of the magnetization configuration of the whole system. In general, this means we are looking for a solution of the static equation (3.15) and therefore do not need an additional time integration. However, from an applied point of view, trying to solve the static equation of

5.1 Weak formulation

micromagnetism directly by finite element methods does not usually succeed because the linear system solver does not converge. The reason can be found in a "bad" initial guess that is too far away from the actual solution. In order to solve only the static systems $Ax = b$, a "good" initial guess is necessary. In our approach, such an initial guess is obtained by solving the dynamic equations for a specified number of time steps. Since we are not actually interested in the evolution, the damping parameter α can be adjusted to decrease the solving time. Motivated by our findings in section 4.4, we usually set $\alpha = 1$ for preconditioning purposes. Such a choice leads to strong reduction of oscillations within the solution. After a certain calculation time t_{stop}, the time dependent solver is aborted and the stationary solution is solved employing $u_h(t_{\text{stop}})$ as an initial guess.

For the numerical solving process, certain scales are of importance and govern the discretization of the magnetic domain Ω_{mag}. In respect to the material parameters, a typical length scale L is given by (Hubert and Schäfer, 2000)

$$L = \sqrt{\frac{A}{K}} \qquad (5.8)$$

with the exchange constant A and the energy scale K. This length scale may be taken as a measure of resolution of the finite element mesh. The employed domain discretization needs to be sufficient to resolve structures of at least the dimensions L.

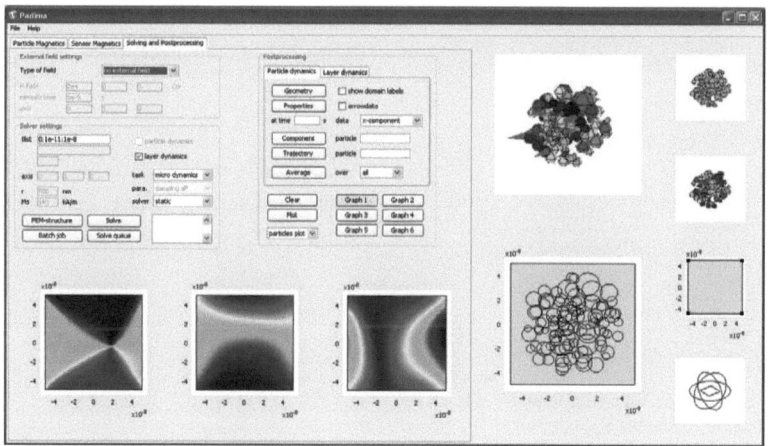

Figure 5.4: Graphical user interface of PADIMA used for model setup enabling various applications in COMSOL Multiphysics.

5. Particle detection

In the framework of this thesis, an easy to handle COMSOL plug-in, PADIMA, was developed to offer many of the applications described in the following sections in a predefined form and therefore requires the user to specify only a small set of parameters for the model setup. A short introduction is given in Appendix A.4

5.2 Manipulation of magnetic vortex states

To test the implementation into COMSOL Multiphysiscs™, we applied our model to the following situation: A circular magnetic layer of diameter d and thickness h as shown in Figure 5.5 goes over to a vortex state. As a second magnetic layer, we consider a homogeneously magnetized electrode with magnetization oriented parallel to the y-axis. The vortex may be manipulated by an external homogeneous magnetic field leading to different TMR-ratios if we assume a second magnetic layer with a homogeneous magnetization distribution parallel to the y-axis. For the sake of simplicity, we will only consider the free layer. As layer material, we assume CoFeB in an amorphous phase, leading to

$$A_{\text{CoFeB}} = 2.86 \cdot 10^{-11} \text{J/m}$$
$$M_{S,\text{CoFeB}} = 1194 \text{kA/m}^2$$

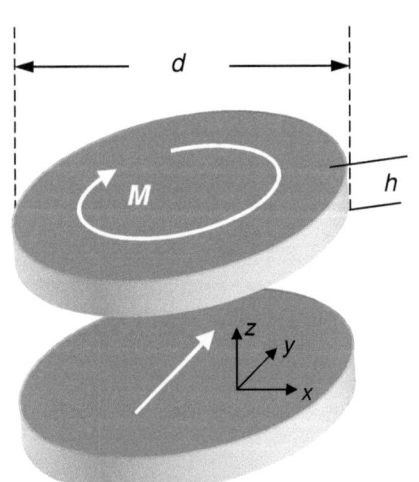

Figure 5.5: Vortex state in a circular disc of diameter d and thickness h.

according to (Blizer et al., 2006) and $f_{\text{ani}} \equiv 0$. Experimental investigations on systems of this type can be found in the works of R. Lehndorff et al., 2007; 2008; 2009. Applying a homogeneous, external magnetic field H, the vortex position changes. Analytic discussions of these systems by K.Y. Guslienko et al., 2001; 2006; 2008 predict a response of type

$$\frac{h}{d} \approx \left(\frac{d\text{TMR}}{dH} \right)^{-1} \bigg|_{H=0} \qquad (5.9)$$

5.2 Magnetic vortex states

Equation (5.3) has been solved, choosing an undisturbed vortex state as the initial configuration. Simulations have been carried out for the parameters $d = 300$, 500, and 700 nm and $h = 10$, 15, and 20 nm. If an external homogeneous magnetic field along the x-direction is applied, the vortex core changes its position. The displacement occurs into the direction of the region, where external field and magnetization distribution are antiparallel to each other. Thus, the system minimizes its Zeeman energy. The position of the vortex in dependency of the external magnetic field is shown in the magnetization plots of Figure 5.6 for the case of $d = 500$ nm and $h = 20$ nm. For small fields, this shift is proportional to the external field. If the vortex reaches the boundary of the disc, an increase of the magnetic exchange energy

$$E_A = \int_{\Omega_{mag}} \frac{2A}{\mu_0 M_s} (\nabla \hat{m})^2 dx$$

with Ω_{mag} the magnetic domain, can be observed (Figure 5.6, top plot). This leads to a motion of the vortex. Any further increase of the external field results in a complete alignment of the soft electrode with the external field direction; the vortex is driven out of the layer. From the calculated magnetization patterns, the TMR-ratio can be obtained

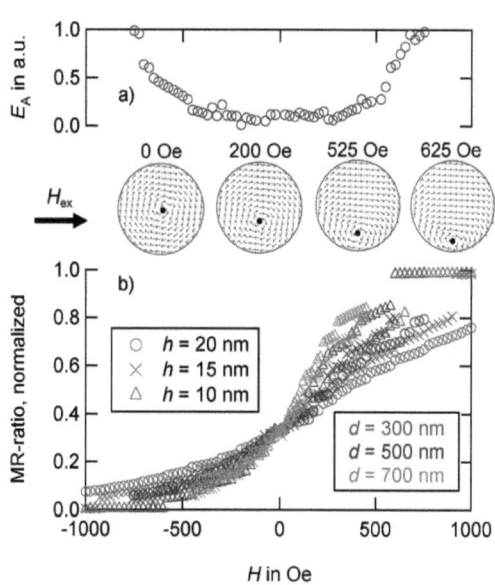

Figure 5.6: (a) Influence of the magnetic field on the vortex position and the exchange energy E_A in respect to the applied field. (b) resulting TMR-ratios for different d and h.

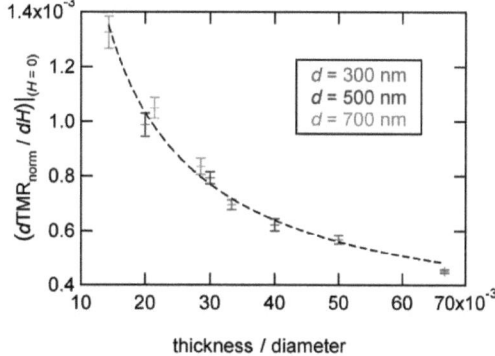

Figure 5.7: Influence of the ratio between thickness and diameter on the derivative $(d\,TMR/dH)_{H=0}$. The dashed line is a fit of $a/TMR + b$ to the obtained data.

119

5. Particle detection

evaluating the angle between the magnetization in the upper and the lower electrode at each point of the sensor according to (5.8). Simulations also show that the TMR-ratio remains constant if the magnetization orientation in the lower electrode and the external field are orthogonal to each other. Figure 5.7 is obtained by plotting $(d\text{TMR}/dH)_{H=0}$ against the ratio h/d. In particular, we find the analytically predicted behaviour.

5.3 Space resolutive magnetic detection: "magnetic lenses"

As already discussed in section 4.3, a homogeneously magnetized sphere creates a magnetic dipolar stray field in the entire space. Close to the particle itself, such a field is rather inhomogeneous. In particular, we may not assume a homogeneous particle contribution along the sensor area if the particle, sensor and the distance between the particle and sensor are on the same size scale. Based on this observation, we may also conclude that a change of the particle position entails a change of the magnetic field along the sensitive area and, thus, also a change of the sensor response according to section 5.1.1. If it would be possible to find a correlation between the TMR-response and the position of the particle, a measurement would provide additional information instead of only the answer to a yes/no-question.

The goal of this section is to estimate the limits for a magnetoresistive detection of magnetic particles under the constraint that we want to predict the position of the particle as precisely as possible. In all our considerations we will assume CoFeB-layers with a magnetization of $M_S = 1194$ kA/m and an exchange constant of $A = 2.86 \cdot 10^{-11}$ J/m (Blizer et al., 2006). For the sake of simplicity, we will also assume CoFeB in an amorphous phase by setting $f_{\text{ani}} \equiv 0$. However, before we can employ our model to make predictions, a comparison between theoretical calculations and experimentally obtained data is necessary in order to ensure our analysis resemble the actual situation.

5.3.1 Comparison to experimental data

As reference data, we choose the experimental situation studied in the PhD theses of C. Albon, 2009: Magnetic tunnel junctions (MTJ) with the following configuration

5.3 Detection properties

(thicknesses given in nanometer) was developed: Ta(6.5) / Cu(30) / Ta(19) / Cu(9) / MnIr(12) / CoFe(3) / Ru(0.9) / CoFeB(2.8) / MgO(1.5) / CoFeB(4) / Ru(8) /Ta(4) / Au(50), compare Figure 5.8. The MTJ was created on a thermally oxidized Si/SiO$_2$-wafer using DC-magnetron sputtering for metal targets and RF-magnetron sputtering for the MgO barrier. The shape of the sensor was chosen elliptical with a length of 400 nm on the longitudinal and 100 nm on the transverse axis, respectively. The experimental setup and its properties are shown in Figure 5.9.

For the discussions here only the following details are of concern: The bottom ferromagnetic layer was pinned by annealing the stack in a vacuum at 350°C for one

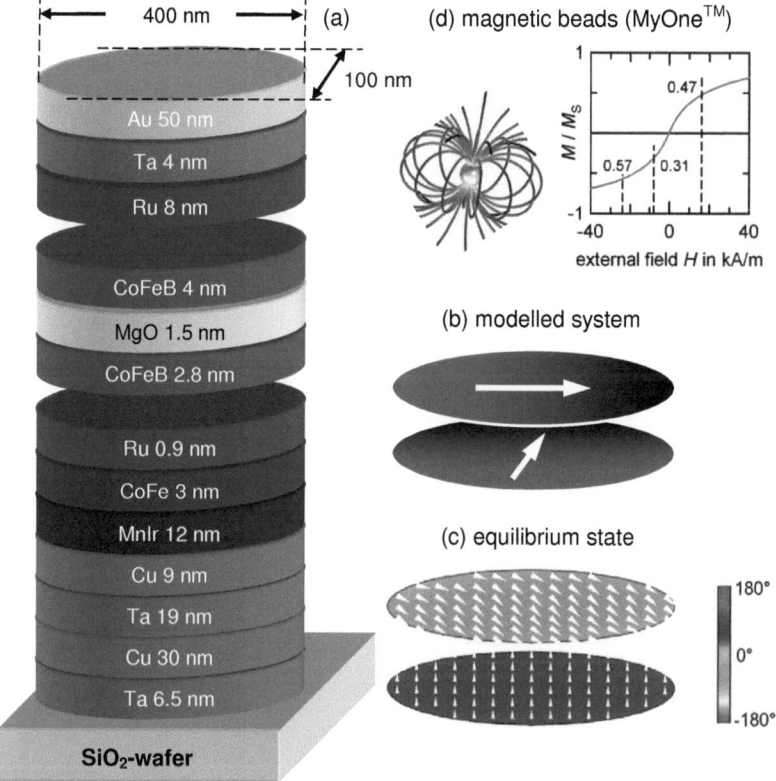

Figure 5.8: Layer configuration employed in the experiments. (a) For the theoretical simulations it is sufficient to focus on the CoFeB / MgO / CoFeB-subsystem. (b) Magnetization configuration of the magnetic trilayer system and (c) equilibrium state obtained as a solution of the micromagnetic equations if there is no external field applied. The bottom electrode is assumed to be pinned; the top is free and attains its free state due to the interplay of geometrical anisotropy and stray field coupling to the lower ferromagnetic film. (d) magnetic characterization of the employed MyOne-beads, obtained from AGM-measurements.

5. Particle detection

hour in an external magnetic field of 6000 Oe applied parallel to the hard (transversal) axis of the element. Due to high shape anisotropy of the elements, the magnetization of the soft ferromagnetic electrode aligns parallel to the long axis of the ellipse (compare also section 3.2). This setup enforces a linear output of the MTJ-element over a magnetic field range of ±500 Oe. The possibility of such strong fields allows for bringing the particles close to saturation which in particular makes TMR-sensors suitable for a direct detection of magnetic markers with no further electronic amplifications.

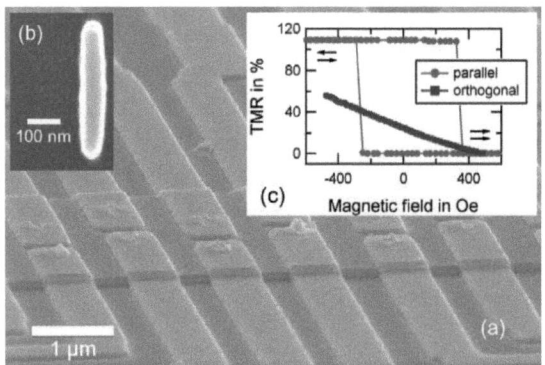

Figure 5.9: (a) Experimental realization of the sensor array showing the sensor elements and the connections to the outside by conducting lines, (b) single sensor element, (c) representation of the sensor output having two ferromagnetic electrodes in a parallel or an orthogonal orientation.

Similar to the experimental realizations of other devices discussed superparamagnetic 1 μm Dynabeads MyOne^{TM} were used as magnetic markers and placed on top of the sensor by a dropping procedure. For the actual detection, an in-plane DC magnetic field was applied over the linear range of the sensor. Since the sensitivity varies between each sensor, we introduce the relative resistance change via

$$\Delta \text{TMR} = \frac{R_{part} - R_{sensor}}{R_{sensor}}$$

with R_{part} and R_{sensor} the resistance of the TMR-sensor with and without a particle, respectively. If we apply (5.5) and refer to TMR$_{part}$ and TMR$_{sensor}$ for the TMR-value with and without a particle, respectively, we may rewrite the definition of ΔTMR according to

$$\Delta \text{TMR} = \frac{R_{part} - R_{sensor}}{R_{sensor}} = \frac{(R_{part} - R_0) - (R_{sensor} - R_0)}{R_{sensor}}$$
$$= \frac{(R_{part} - R_0)/R_0 - (R_{sensor} - R_0)/R_0}{R_{sensor}/R_0}$$

5.3 Detection properties

$$= \frac{(R_{part} - R_0)/R_0 - (R_{sensor} - R_0)/R_0}{(R_{sensor} - R_0)/R_0 + 1}$$

$$= \frac{TMR_{part} - TMR_{sensor}}{TMR_{sensor} + 1} \quad (5.10)$$

with R_0 the resistance value of the free sensor. Equation (5.10) will be used to compare different sensors with each other. In order to investigate the response of the sensor in respect to the particle position, we introduce a discrete grid above the sensing free layer. The grid nodes r_{part} are given by

$$\begin{aligned} x_{part} &= -1\,\mu m + (i-1) \cdot 0.2\,\mu m, & i &= 1, \ldots, 11 \\ y_{part} &= -1\,\mu m + (j-1) \cdot 0.1\,\mu m, & j &= 1, \ldots, 21 \\ z_{part} &= 0.562\,\mu m \end{aligned} \quad (5.11)$$

The choice of the z-coordinate results since no passivation layer can be found on the sensor surface. Therefore, a particle deposited directly on the surface has a distance from the soft layer given by the sum of particle radius and the thicknesses of all layers in between, in detail:

$$z_{part} = 0.5\,\mu m + 8\,nm(Ru) + 4\,nm(Ta) + 50\,nm(Au) = 0.562\,\mu m$$

Due to the pinning of the bottom CoFeB-layer, we only take the magnetization direction $\hat{m}_1$ as free, the distribution $\hat{m}_2$ in the lower layer is set to $\hat{m}_2 = \hat{y}$. The two magnetic layers are coupled via their stray field. Additionally, we assume a Néel-coupling according to (5.7) with structural parameters $\lambda = 30$ nm and $h = 3$ Å. The equilibrium state is presented in Figure 5.8(c). The stray field coupling is apparent: instead of

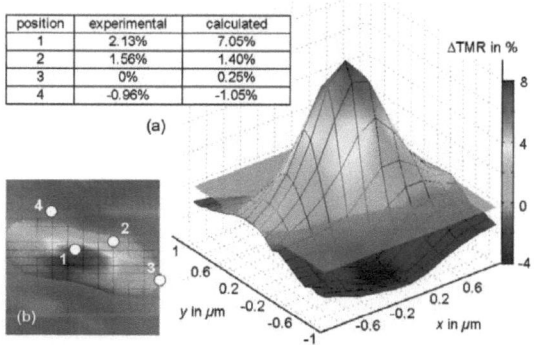

position	experimental	calculated
1	2.13%	7.05%
2	1.56%	1.40%
3	0%	0.25%
4	-0.96%	-1.05%

Figure 5.10: Comparison between experimental data and theoretical calculations. (a) ΔTMR-values for particle positions at the grid nodes. The grey level separates positive and negative values. (b) Top view of the ΔTMR-map, white markers indicate the bead positions for the comparison.

123

5. Particle detection

aligning parallel to the soft geometrical axis of the ellipse, the magnetization vector is rotated counter clockwise attaining a position between geometrical easy and hard axes. For the calculations, we consider an external magnetic field parallel to the y-axis with $H_y = 24$ kA/m which corresponds to a particle saturation of 0.56 times the saturation magnetization according Figure 5.8(d). This corresponds to a magnetization of 67.2 kA/m if Dynabeads® MyOne™ are considered (Fonnum et al., 2005).

Calculating the magnetization distribution in the free layer with a particle at the position given by (5.11), we obtain discrete ΔTMR-values that have been extended by linear interpolation to the ΔTMR-map presented in Figure 5.10(a). Comparing the calculation results with the experimental findings, a very strong quantitative agreement may be reported. The deviation of the ratio at the centre position can be explained by local topographical changes which have strong influences on the measured signal if the bead is close to the sensor.

5.3.2 Estimation of the spatial resolution limits

Due to the very strong agreement between experimental data and the simulation results, the model introduced may be used to analyze the limitations when estimating the particle position in respect to the sensor. Figure 5.11 shows again a top view of a TMR-map for an external field $H_y = 16$ kA/m. The response map shows a maximum value TMR$_{max}$ close to the sensor position and two local minima TMR$_{min}$ in some distance from it. The response is symme-tric to two axes which are obtained by a rotation of the ellipse axes of an angle α. This rotation originates from the coupling of the two magnetic layers. Sign and size of α depend on the interplay of the two coupling effects mentioned above: While Néel-coupling favours parallel alignment, the stray field interaction between the two layers lead to

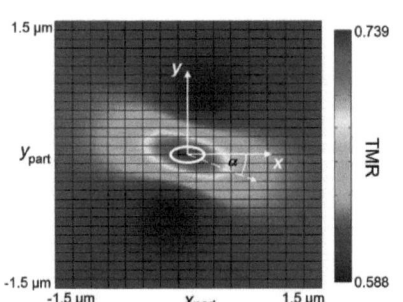

Figure 5.11: TMR-map for a homogeneous magnetic field of 16 kA/m along the y-axis. Crossings of black lines correspond to grid nodes; the origin of the coordinate system coincides with the centre of the sensor (white ellipse). The pattern shows symmetries, though the symmetry axes do not coincide with the semiaxes of the ellipse but are turned by an angle α due to the coupling (stray field/Néel-coupling) of the electrodes.

5.3 Detection properties

an antiparallel configuration. For sensors of an area smaller than 1 μm², stray field interaction is commonly dominating (Meyners, 2006).

In a first step, we analyze the influence of the external magnetic field on the sensor response. Similar calculations as presented in Figure 5.11 are carried out for field values of H_y = 8 kA/m, 24 kA/m, 40 kA/m and 56 kA/m. The particle magnetization is chosen according to Figure 5.8(d) and given by 0.31, 0.57, 0.69 and 0.76 times the saturation value, respectively. To ease the comparison between different maps, we employ again the relative ΔTMR-ratio

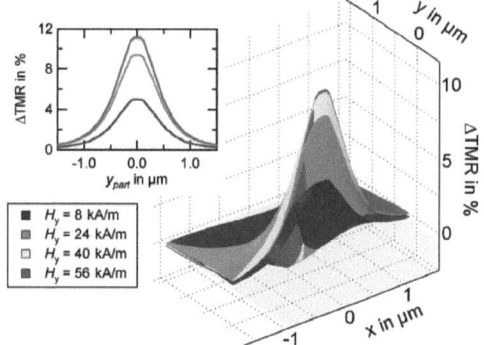

Figure 5.12: ΔTMR-maps for z_{part} = 0.562 μm and external field strengths of H = 8, 24, 40 and 56 kA/m applied parallel to the y-axis. The degree of particle saturation is given by 0.31, 0.57, 0.69 and 0.76. The inset shows crosscuts through the centre along the y-axis.

$$\Delta\text{TMR} = \frac{\text{TMR}_{part} - \text{TMR}_{stack}}{\text{TMR}_{stack} + 1} \quad (5.10)$$

The results are shown in Figure 5.12. We find that an increasing external field value leads to an increasing effect change at the centre of the sensor. However at the same time, the measurable TMR_{part}-values decrease as

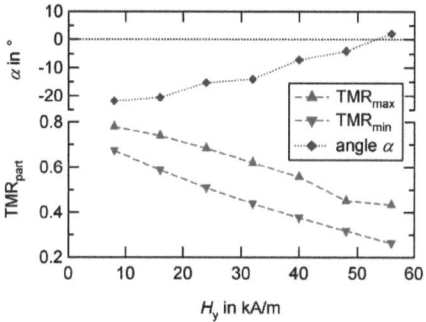

Figure 5.13: Sensor characteristics for different external fields applied parallel to the y-axis. For high fields the maximum and minimum measurable TMR_{part}-values decrease. The angle α increases and reaches zero close to H_y = 56 kA/m.

shown in Figure 5.12. In particular, they will rapidly drop below the threshold of noise that can always be found in such devices (Hedwig, 2009). Therefore, high fields increase the resolution of the sensor, but decrease the area in which a particle can be detected. This is similar to the behaviour of optical lenses. If a critical field value is exceeded, the response of the sensor changes (Figure 5.12, H_y = 56 kA/m); the maximum ΔTMR-value does not increase any further but decreases together with TMR_{part}.

5. Particle detection

An analysis of the angle values shows that this point coincides with $\alpha = 0$ (Figure 5.13). This behaviour is due to two factors: while the external magnetic field increases linearly, the particle moment reaches saturation and therefore, the external field will overcome the particle influence at a certain point. On the other hand, the sensing free top layer reaches saturation and becomes less sensitive. Thus, it should be pointed out that ideal detection conditions can be obtained by adjusting the external magnetic field to the particles to be detected and the sensor chosen for the detection.

The obtained TMR-maps can be used to estimate the position of the magnetic particle. Here, we will focus on the distance d

$$d = \sqrt{x_{part}^2 + y_{part}^2} \qquad \text{for} \qquad z_{part} = 0.562 \mu m \qquad (5.11)$$

between the particle and the ellipse centre of the top electrode. As is already apparent from Figure 5.11, a single TMR-value corresponds to several particle positions and can therefore only give an upper and a lower bound for the distance. To analyze these bounds, we divide the interval [TMR_{min}, TMR_{max}] into N equally sized sub-intervals $\{\Delta MR_i\}_{i=1,...,N}$. For our analysis, the size of ΔMR_i is basically arbitrary; it only needs to introduce a proper class division of the data points along the discrete grid nodes. For the data analysis we choose $N = 100$. In the experimental situation however, the minimal size of ΔMR_i will correspond to the achievable exactness of the measurement. The map is divided according to the assignment of each map point to the corresponding interval ΔMR_i. This defines a relation between ΔMR_i and the distance d of the regarded map

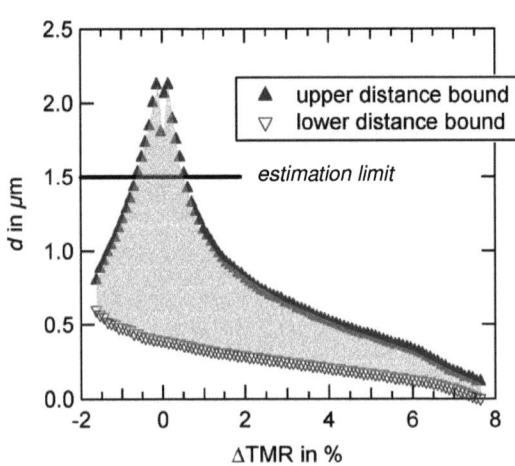

Figure 5.14: Upper and lower distance bounds obtained from the map shown in Figure 5.11. According to a measured ΔTMR-value, the distance can be estimated by the highlighted area. The estimation limit results from the finite grid size of 3 μm. For ΔTMR-values found along the grid boundary no evaluation is possible.

5.3 Detection properties

point to the centre of the sensor. A $\Delta MR_i - d$ -plot is presented in Figure 5.14. Due to the construction each, ΔMR_i -interval corresponds to a set of d-values. A single measurement can therefore only give an upper and a lower bound for the distance between particle and sensor; the two lines of data points show these bounds. At a given ΔTMR-value, each distance value in between is possible. However if additional measuring directions are taken into account, further information can be obtained.

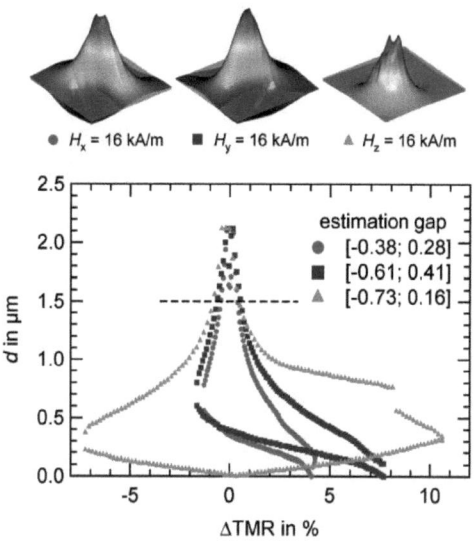

Figure 5.15: Plots for upper and lower d-bounds obtained form the TMR-maps for external magnetic fields of 16 kA/m along the positive coordinate axes. The intervals given, the estimations gaps, indicate the range of ΔTMR where there is no estimation possible because not all corresponding values con be found on the grid chosen. Jumps in the ΔTMR surface can be attributed to a finite mesh resolution.

Referring to the data of our calculations, we can estimate the degree of accuracy to which the distance d between particle and sensor centre can be determined by combining different measurement directions. As a simplification, we will assume that the particle is stationary during all measurements. As can be seen from Figure 5.12, employing different field values along the y-axis will cause several difficulties: the ΔTMR-response changes at every position in the same manner, therefore, additional information on the particle position can only originate from a varying angle α. This effect, however, is difficult to exploit as significant angle perturbations require fields lying in the range of particle and sensor saturation. This reduces the visibility field as already explained above by a large amount which leads to a decreased spatial resolution.

Instead, we combine bound estimations for applying external magnetic fields parallel to x- and z-axis. Thus, we obtain upper and lower bounds

$$d_{up}^{H_x}, d_{up}^{H_y}, d_{up}^{H_z} \quad \text{and} \quad d_{down}^{H_x}, d_{down}^{H_y}, d_{down}^{H_z},$$

127

5. Particle detection

respectively, according to Figure 5.15. The best estimation for the distance d is given by the interval

$$I = [d_{\text{down}}, d_{\text{up}}] := \left[\max_{H_x, H_y, H_z} d_{\text{down}}^H, \min_{H_x, H_y, H_z} d_{\text{up}}^H \right] \quad (5.12)$$

where the minimum and the maximum need to be taken over all three measuring directions. For the evaluation, only ΔTMR-values are taken into account which cannot be found along the boundary of the grid. Based on our calculation, no estimation is possible for these; values are given in Figure 5.15. The distance estimates for the combined detection d_{up} and d_{down} are shown in Figure 5.16(a). Upper and lower bound almost coincide near to the sensor, while the interval size increases with increasing particle-sensor distance d. Certain features (local maxima in the lower bound-map, highlighted in Figure 5.16) can be found due to the combined measurement which increases the accuracy of the position detection.

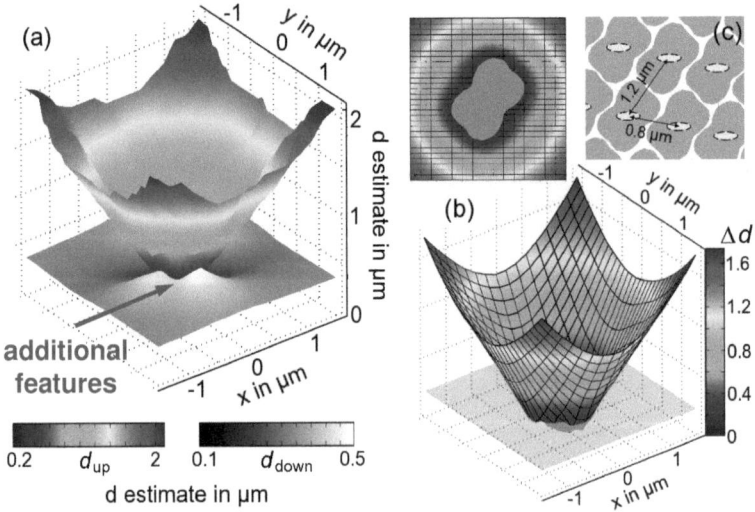

Figure 5.16: Error estimates according to the simulation data of Figure 5.15. (a) shows upper and lower bound for each grid point, a particle at a given coordinate leads to a corresponding distance estimation. Different features can be found in the lower bound d_{down} which originate from the combination of several estimations. (b) presents the maximum error Δd from the actual distance. The grey level coincides with the plane $\Delta d = 0.2 \, \mu$m. The subplot (c) shows a possible sensor array where the sensor positions are chosen according to (b) so that a detection with a precision of $\Delta d = 0.2 \, \mu$m is achieved.

5.3 Detection properties

The maximum absolute error

$$\Delta d = \max_{\delta = \{d_{up}, d_{down}\}} |d - \delta| \qquad (5.13)$$

is presented in Figure 5.16(b). If a certain threshold of minimum exactness is defined, we obtain guidelines on how to construct sensor arrays, which ensure such measurement resolution along all the *x-y*-plane. Figure 5.16(c) shows the top view to the total error surface, the grey area corresponds to estimation bounds below an error of 0.2 μm and the resulting sensor assembly.

5.3.3 Sensors for continuous-flow particle measurements

If particles on the size scale of nanometers are considered, two different problems usually need to be addressed: (1) though the magnetization of the particle material can reach much higher values (Hütten et al., 2005), their over-all magnetic moment is still far smaller due to the particle size. (2) Minimizing the detecting signal sensors leads to a change of the dominating energy contributions: the smaller the sensor dimensions, the more important the interplay between exchange energy and layer coupling via stray field. This leads to a strong confinement of the magnetization distribution, i.e. sensors consist of a signal domain. In particular, they show no hysteresis but

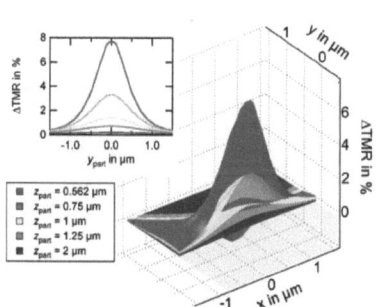

Figure 5.17: ΔTMR-maps for different heights: z_{part} = 0.562, 0.75, 1, 1.25, and 2 μm for H = 16 kA/m applied parallel to the *y*-axis. The two-dimensional plots present cross-sections along $x = 0$.

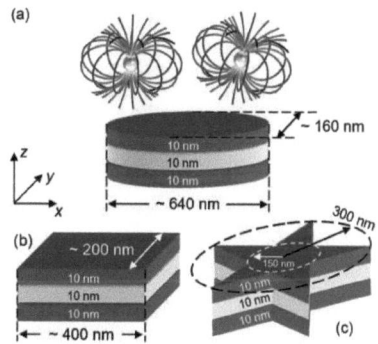

Figure 5.18: Schematic representation of the investigated sensor setups (a) ellipse similar to the preliminary sections, (b) a rectangle and (c) a star.

129

5. Particle detection

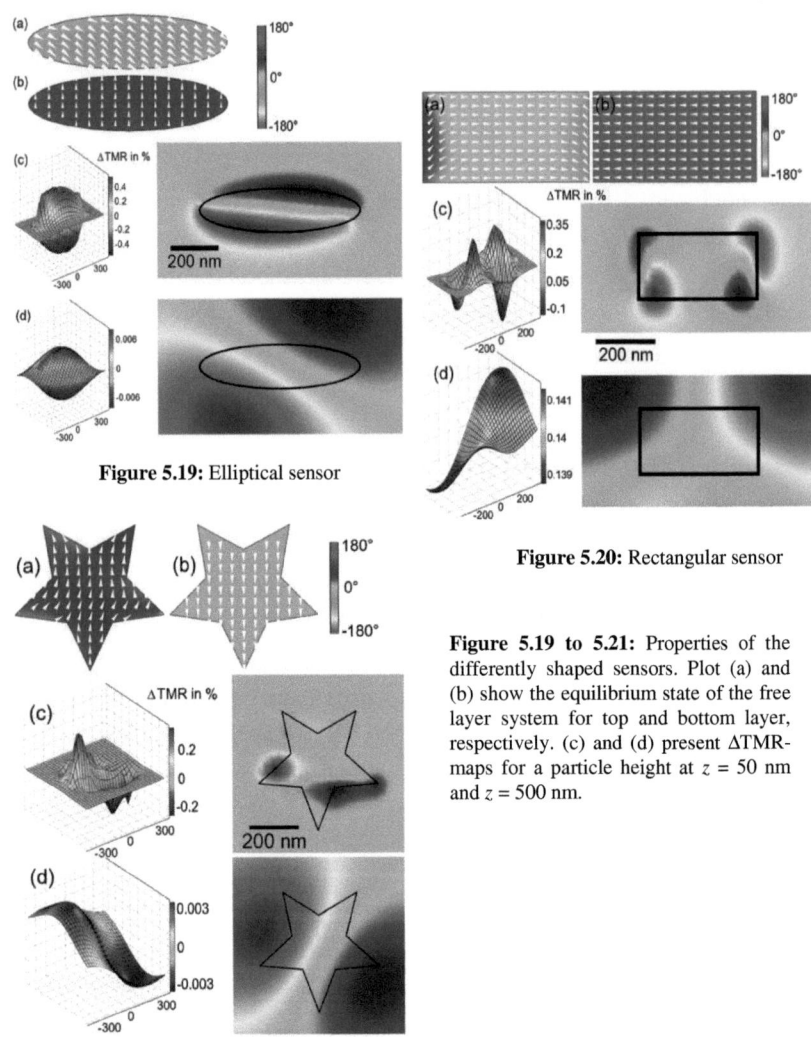

Figure 5.19: Elliptical sensor

Figure 5.20: Rectangular sensor

Figure 5.21: Elliptical sensor

Figure 5.19 to 5.21: Properties of the differently shaped sensors. Plot (a) and (b) show the equilibrium state of the free layer system for top and bottom layer, respectively. (c) and (d) present ΔTMR-maps for a particle height at $z = 50$ nm and $z = 500$ nm.

also behave rather stiffly. A direct consequence of these difficulties can be seen in the error plot of Figure 5.16. At a certain distance, the magnetic particle stray field is no longer sufficient to evoke a strong sensor response. This results in a large error and usually in a signal below the noise threshold. While a lack of detection range parallel to the x-y-plane can be handled by designing a sensor array as discussed in section 5.3.2, there is no such possibility available to increase the view in z-direction. This is

5.3 Detection properties

Table 5.1: Grid parameters for different sensor geometries.

	x_{min} / x_{max}	y_{min} / y_{max}	steps
elliptical	-0.5 / 0.5 µm	-0.25 / 0.25 µm	30 / 30
rectangular	-0.4 / 0.4 µm	-0.2 / 0.2 µm	40 / 20
star-shaped	-0.4 / 0.4 µm	-0.4 / 0.4 µm	30 / 30

also not necessary for our original problem as we consider magnetic markers which immediately bind to the surface. However, recent developments have revealed a great deal of interest for the detection of magnetic beads in continuous flow devices (Loureiro, 2009). Figure 5.17 shows TMR-maps for particles magnetized by an external field of 16 kA/m parallel to the y-axis for different particle heights. The response of the sensor decreases rapidly with distance. To ensure detection a higher field value according to Figure 5.12 might be essential for experimental measurements.

However, this would be undertaken at the cost of visibility field. In this section we pursue a different strategy: the adjustment of the sensor geometry. Three different sensor geometries to compare are shown in Figure 5.18: (a) a sensor of elliptical shape, (b) a rectangular layout and (c) a star-shaped multilayer with S_5-symmetry. The dimensions for the cases (a) and (b) are chosen so that each geometry has an identical area of 80,000 nm². Further, we again assume a TMR-sensor consisting of two ferromagnetic CoFeB-layers separated by an insulating tunnelling barrier. In all configurations the bottom electrode has a fixed magnetization direction. The magnetization direction in the sensing free top electrode is calculated as a solution of the Brown equation (3.15). For the comparison we will neglect Néel coupling, which does not imply a strong restriction as the layer coupling is mainly dominated by the stray field coupling as we have already noticed in section 5.3.1. For the quantification of the sensitivity, a homogeneously magnetized probe particle of radius $R = 20$ nm and saturation magnetization $M_{part} = 1000$ kA/m with magnetization vector perpendicular to the film is placed along a discrete grid with grid nodes at r_{part}. The nodes run equidistantly between the values x_{min}, x_{max} and y_{min}, y_{max}, respectively. Parameters for different geometries are given in Table 5.1.

The dipolar magnetic particle field is calculated by (4.2) which leads to ΔTMR-maps in a manner similar to our approach in the preceding section. The results are

5. Particle detection

summarized in Figures 5.19 to 5.21. The subplots (a) and (b) show the equilibrium configurations of the upper free sensing layer and the pinned bottom electrode, respectively. (c) and (d) present ΔTMR-maps for a distance between the centre of the magnetic particle to the sensing electrode of 50 and 500 nm. All cases have one thing in common: highly sensitive areas can only be found along the boundaries of the electrodes if the particles are near to the sensor (subplots (c)). A particle placed directly on top of the sensor but in some distance to the boundary segments does not contribute to a strong response. This effect is due to the above mentioned stiffness of the single domain elements. Along the centre of the electrode, the magnetic exchange energy contribution prohibits a variation of the magnetization distribution in this regime. Therefore, sensitive areas can only be found along the boundary.

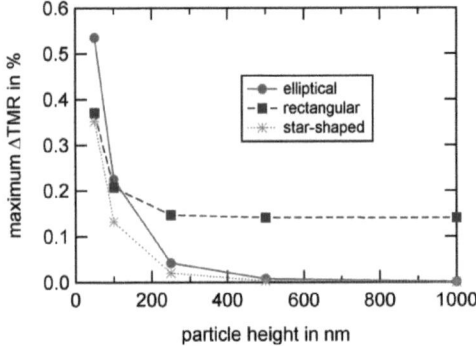

Figure 5.22: Comparison of the maximum sensor response of the investigated sensor geometries. While elliptical and star-shaped geometries cannot detect particles at a certain distance, the rectangular shaped sensor maintains its detecting properties.

In detail, elliptical sensor elements show high response along the whole surface due to its lateral dimensions; particles are always close to a certain geometry edge. Therefore, the maximum signal overall is comparatively high in the case of small distance detections (Figure 5.19). Rectangular geometries show instead only high sensitivity along the short boundary segments, Figure 5.20(c). This behaviour originates from the interplay between demagnetization energy (favouring a magnetization orientation along the x-axis) and the stray field coupling to the bottom layer (leading to a C- or S-state). This is already apparent from the equilibrium state where high curvature of the magnetic components along the short boundaries can be found. These areas can be switched with very low switching field into an orientation antiparallel to the magnetic alignment within the bottom electrode. As we see in Figure 5.22, the maximum absolute change is smaller than the corresponding value for the elliptical geometry as fewer boundaries contribute to the sensor response. However, due to small switching energies, rectangular sensor geometries enable particle detection at much higher distances. According to Figure 5.20(d), the values at the grid nodes do not vary strongly for the high distance case.

Therefore, a space resolutive detection is no longer possible but the setup provides information if a magnetic particle is near.

We may extend this approach by the introduction of sensor layouts with higher geometric complexity. An example is the star-shaped setup (Figure 5.21). Similar to the situation discussed above, we find strong responses along boundary segments where the orientation of the magnetization is not antiparallel to the alignment within the bottom layer. Due to the geometrical features, the ΔTMR-map shows numerous details which should enable a very high spatial resolution. Furthermore, the relative sensitive area is larger when compared to the rectangular setting. However, due to strong geometric anisotropies, these properties cannot be maintained at long distances.

5.4 Number resolutive magnetic detection

Depending on the application, rather than an estimation of the particle position, it might be necessary to obtain information on the number of particles that are placed on the sensor surface. Employing a magnetoresistive sensor with a sensor response independent in respect to the particle position, it is already well established by the group of S. Sun (e.g. Shen et al., 2008a, b) that the sensor response is linearly connected to the number of particles in detection range.

In order to meet the requirement of a position-independent response, the particle size should be far smaller than the sensor dimensions. For out discussion, we choose the same setup as discussed in section 5.3 (compare Figure 5.8). In the experiments that were carried out by C. Albon, 2009, 14 nm Co-nanoparticles were placed by a dropping procedure on top of the sensor. A typical experimental situation is shown in Figure 5.23. We shall briefly summarize the experimental observations: If we employ again the relative change ΔTMR, the measured range can be divided into three different regimes:

1) **Very low coverage:** With the setup described above it was possible to detect 15 dispersed nanoparticles. Below this threshold, the obtained signal is below the detection threshold of the device as the sensor response is dominated by noise.
2) **Dispersed particles:** No hysteresis can be found in the measured signal. The distance between particles is large enough that their coupling strength may be omitted.

5. Particle detection

3) **Interaction regime:** Due to high surface coverage, the average distance between adjacent particles is small; dipolar coupling (see chapter 4) becomes important for the magnetic state of the particles. A hysteresis can be found showing a coercivity which coincides with the coercitive field of the particles.

The three regimes can be found in Figure 5.24.

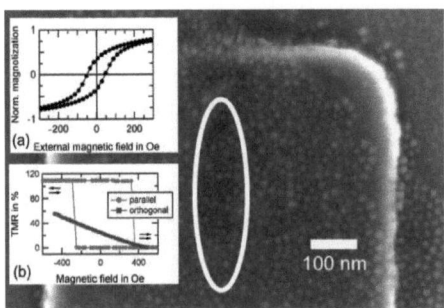

Figure 5.23: Experimental setup for the multi-particle detection. 14 nm Co-particles are place by a dropping procedure on top of the sensor surface. (a) Hysteresis behaviour of the particles (from AGM measurements) and (b) sensor properties.

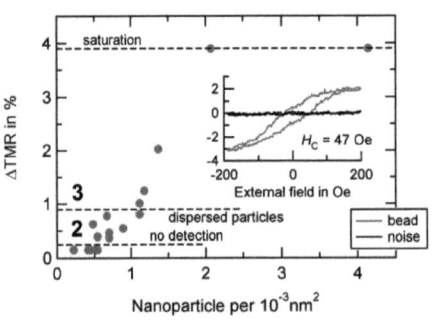

Figure 5.24: ΔTMR in dependency of the sensor coverage strong by magnetic nanoparticles. The measurements can be divided in three regimes. For high particle coverage strong dipolar coupling can be observed leading to a hysteresis in the obtained sensor signal (inlet).

It is already well established (Wang and Li, 2008) that the sensor response is linearly connected to the number of particles in detection range if particle-particle interactions can be omitted and if all positions contribute in the same way. Since simple models calculating only the magnetic field along the sensor surface and retrieving the TMR-effect by field integration lead to a qualitative agreement with experimental situations for very simple systems, this can be generalized. Non-interacting magnetic particles contribute linearly, with the contribution weighted by the single particle effect.

In order to understand the coupling phenomena observed in the experiments and shown in Figure 5.24, we apply the layer model introduced in section 5.1 and combine it with the interacting particle calculations discussed in chapter 4. Therefore, we introduce a monolayer of magnetic nanoparticles on top of the sensor. We assume particles to be assembled in a two-dimensional hexagonal particle lattice, as shown in Figure 5.25. To change the degree of coverage, the distance between neighbouring particles is varied. Particles are placed at a height $z_{part} = 62$ nm + particle radius, according to the layer stack employed for the experimental realiza-

5.4 Number sensitive detection

tion. The saturation magnetization of the particles is set to the bulk value of Co. The results of the calculations are shown in Figure 5.26. For low coverage, we find a linear dependency between ΔTMR and the degree of coverage. Furthermore, we find a coupling strength that is sufficiently small to be overcome by thermal contributions. The findings correspond to the free particle case, i.e. the first and the second regime referring to the experimental findings. The response increases linearly with the number of particles. For high coverage (> $6 \cdot 10^{-4}$ nanoparticles per nm^2) saturation similar to the experimental findings can be observed. In this regime, the calculated signal shows hysteresis (Figure 5.26, inlet).

Qualitatively, the numerical calculations fit the experimental findings very well. The sensor response is linear in respect to the number of particles as long as dispersed particles are considered, i.e. the dipolar coupling strength is sufficiently low. In this range, the sensor is therefore suitable for applications seeking number sensitive detection. Additionally, in correspondence to the experimental data, our approach introducing a dipole-dipole coupling between particles, leads to a saturation of the measured signal for high coverage and an additional hysteresis within the sensor response. We may therefore conclude that the experimental observations are a direct proof for the dipolar coupling between magnetic nanoparticles.

However, if we compare the actual numbers, we find deviations that may be attributed to different causes: the saturation magnetization for the particles might be too high. Generally, particle values differ from bulk values. Additionally, fast oxidation of particles

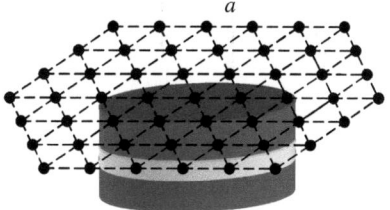

Figure 5.25: Situation in the theoretical model. Particles are placed at nodes of a hexagonal grid with side length a. In contrast to the maps discussed in the preliminary sections, all grid positions are occupied at the same time. Calculations are carried out for different grid parameters, i.e. different surface coverage.

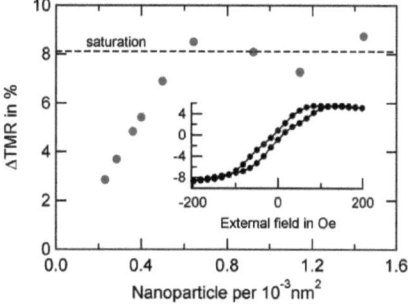

Figure 5.26: Calculated ΔTMR-values for different sensor coverage. The inlet shows the hysteresis of the particles lattice for a high degree of coverage of the sensor surface.

5. Particle detection

leads to a smaller value in the experiments (Chernavskii et al., 2007; Ennen, 2008). Furthermore, the sample preparation does not ensure a single layer of particles on top of the sensor which was the assumption employed by the theoretical model. Moreover, the experimental data find hysteresis within the signal already along the linear area. The reason for this can be found in the very high spatial ordering that was assumed for theoretical modelling. In the same way as particles dissolved in a liquid cluster when brought into an external magnetic field (compare section 4.5, 4.6.1), particles agglomerate under the forces exerted towards each other. In other words, in the experiments a higher degree of coverage corresponds to a higher probability of clusters in the experiments.

5.5 Conclusion and Outlook

In this chapter, we have developed a model for the simulation of a magnetoresistive sensor, based on the Landau-Lifshitz equation. We decomposed the complete equation system by using a three-dimensional approach for the calculation of the layer stray field and recast the calculation of the magnetization into a two-dimensional frame employing different projection mappings. The proposed model is capable to find well-known solution of micromagnetic problems and well-suited for the calculations of small thin film systems below an area of 1 μm^2 in which stray field coupling between neighbouring layers is the main driving force of the system.

In section 5.3, we tested our approach by comparing numerical data with experimental findings on elliptical double-layer TMR-sensors consisting of two ferromagnetic CoFeB-electrodes and could report a very strong agreement (section 5.3.1). Deviations could be explained by topological features in the experimental setup. Since theoretical and experimental data matched very well, we could apply our model to obtain guidelines for the construction of sensor arrays. If we assume a particle that does not change its position during the detection process, we have shown that the distance along the sensor plane between particle and the centre of the sensor can be estimated up to an error smaller than 0.2 µm in a region of a diameter of 1.2 µm around the ellipse with semiaxes of 200 and 50 nm. The measuring idea can easily be extended to the determination of the spatial coordinates of the particle. It was also found that high knowledge of the regarded system is necessary for high resolution. In detail, we found a rotation of the TMR-symmetry to the coordinate axes due to layer cou-

5.5 Conclusion and Outlook

pling via the layer stray fields which cannot be omitted on the size scale of sensors investigated here. It is further possible to adjust the measurement resolution by the strength of the applied external field. In general, high fields enable a high spatial resolution but only along a small visibility field. This observation can help to adjust the setup to different measuring tasks: If the question is to decide whether there is a particle at all somewhere in the range of the sensor, small fields should be applied. If additional information on its position are required, the measuring field needs to be increased. The limit for maximum applied fields is given by the ending of the linear response range.

Since the assembly of sensors in sensor arrays can only increase the detection capabilities within the sensor plane, a different approach needs to be pursued to obtain a high detection range along the perpendicular axis. We have shown that adjusting the sensor shape may lead to enhanced detection properties of tunnel magnetoresisive sensors. Introducing soft magnetic areas along boundary segments may significantly lower the limits of what can still be detected. Therefore, the proposed strategies show guidelines for designing new sensor layouts to detect also single magnetic nanoparticles or to enhance the detection capability for continuous flow detections.

In the final section, the influence of a large number of particles on the sensor was investigated. Similar to results of other groups, simulations predict a linear increase of the sensor response in respect to the number of particles deposited on the sensor surface. In this regime, a number sensitive detection would be possible as long as particles do not agglomerate. In accordance to the experimental observations, no further signal increase can be found above a certain coverage value. Instead, particles couple via their dipolar stray field which entails a hysteresis in the measured sensor signal. Therefore, by comparing experimental observations and theoretical calculations, we were able to prove that the experimental findings are a direct observation of the dipolar particle coupling.

As a final remark we may point out, that the results of this chapter indicate that it might not be possible to design sensors that can maintain all measuring tasks, i.e. (a) have a high space resolution, (b) are number sensitive and (c) may detect particles with a high distance to the sensor. In this chapter, we have given an example of a sensor on the nanoscale for all three individual tasks. However, we have also seen that the sensor properties are hard to combine: (a) and (b) exclude each other since a high space resolution requires the sensor to be on the same size scale as or smaller than the object to detect/analyze whereas in number sensitive detections the sensor should be

5. Particle detection

large in respect to the particles. We realized (c) by introducing areas along the sensor which are magnetically very sensitive and may thus be easily switched. As we have seen, this switching behaviour allows for a high detection range, but not for a high space resolution.

Chapter 6
A MrBead-summary

At this point, we have basically solved all the questions that were raised in the introduction. We have theoretically designed a microfluidic lab-on-a-chip device which can handle all the procedures necessary for the specific detection and were able to experimentally prove its functionality. The experimental setup for the ramp segment developed by several MrBead-partners is shown in Figure 6.1. The MrBead-project has proceeded and reached the stage of prototype development. Patencies on several of the strategies developed have been claimed on the international level and also companies interested in the final product have been found. However, whether a final product will be commercially available in several years is yet unclear. Though the conception seems to work, still a lot of difficulties in the manufacturing processes have to be overcome concerning e.g. electric contacts, leakproof channel setups etc. As was already mentioned in the introduction, when the project was started in 2006, also Philipps worked on a similar approach. A final product is now available; however, the detection is no longer done by magnetoresistive sensors. To my knowledge, it was difficulties in the manufacturing process that led to a change of strategies. Therefore, it will be interesting to see, how the MrBead-approach will (or will not) continue within the next years.

However, I think in the framework of this project, we have learnt a lot about the ideas and challenges of microfluidic lab-on-a-chip systems. I am aware of the fact that many, many more parameters can be changed and discussed in respect to every

6. A MrBead-summary

component introduced in this thesis and there are most probably still a large amount of optimization strategies. In my opinion, though, we have increased the knowledge and the understanding of several important processes. This is especially the case in regard to how to design magnetoresistive sensors/sensor arrays on the nanoscale as well as for the uncoupling of the particle flux from the hydrodynamic flow by directly employing the dipolar particle-particle interaction. At this point, I am very sure that both topics will be refined in the near future and lead to the development of new applications/designs in the field of μTAS-devices.

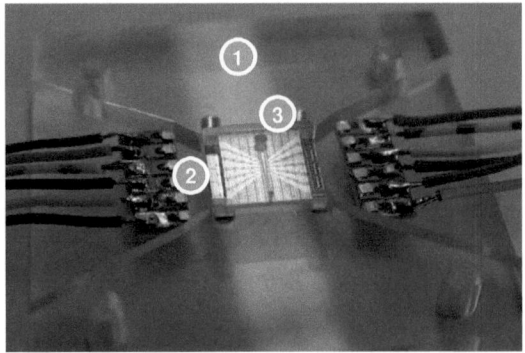

Figure 6.1: Microfluidic ramp structure with sensor array. The carrier (1) has been created by the University of Bielefeld after a design developed by I-Sys [ISys]. The employed sensor array (2) was designed and created by Sensitec [Sensitec]. The microfluidic structure (3) was designed by the University of Bielefeld (section 2.4) and produced by Reiner [Reiner], inlets have been placed by I-Sys. The configuration of the whole setup was realized by the University of Bielefeld, and was mainly carried out by D. Akemeier.

Appendix

A.1 COMSOL Multiphysics™

All calculations presented in this work have been carried out with COMSOL Multiphysics™ which is a commercially available finite element package by COMSOL GmBH, 2005. The package solves time dependent systems which can be written in the form

$$0 = L\left(U, \frac{\partial U}{\partial t}, \frac{\partial^2 U}{\partial t^2}, t\right) - N(U,t)^T \Lambda \qquad (A.1)$$
$$0 = M(U,t)$$

denoting by U the *solution vector*, L the *residual vector*, Λ the *Lagrange multiplier*, N the Jacobian for the constraints and M the *constraint residual vector*. For an exact definition of all components, refer to (COMSOL, 2005).

Before solving the equation system, the applied algorithm eliminates the Lagrange multipliers Λ, in case of linear, time independent constraints M, these are also eliminated. Otherwise, they remain and the original system becomes a differential-algebraic system (DAE). Such a system is solved by the DAE-solver DASPK, 1994, developed by Linda Petzold, University of California, Santa Barbara. The DASPK-solver bases on the DASSL-code, 1996, which employs a backward differential formula of variable step size and order for the solution of the system. The applied numerical integration scheme is implicit which makes it necessary to solve a non-linear equation system in every time step. This is done by Newton-Iteration where the resulting system can be solved by various COMSOL Multiphysics™-solver (see below). The linearization of the above system (A.1) is given by

$$E\frac{d^2 V}{dt^2} + D\frac{dV}{dt} + KV = L - N^T \Lambda \qquad (A.2)$$
$$NV = M$$

with $K = -\dfrac{\partial L}{\partial U}$ the *stiffness matrix*

$D = -\dfrac{\partial L}{\partial(\frac{\partial}{\partial t}U)}$ the *damping matrix*

Appendix

$$E = -\frac{\partial L}{\partial(\frac{\partial^2}{\partial t^2}U)} \qquad \text{the } \textit{mass matrix}$$

For the solution of the linear system, different direct and indirect solvers are possible. In the framework of this work, we mainly used the UMFPACK-solver which is usually applied for large, sparse system matrices. COMSOL Multiphysics employs UMFPACK 4.2 by Timothy A. Davis (UMFPACK, 2009). The system matrix is decomposed by a direct *LU*-decomposition via the *unsymmetric-pattern multifrontal-method* (Davis, 1997; 1999).

COMSOL Multiphysics™ provides a graphical user interface to ease the model setup. However, for the systems discussed in this work, a script language was used (either MatLab or the no longer available COMSOL Script) since the direct implementation is not possible due to the large amount of characters. We will give a short example of how weak equations need to be entered via weak form modelling. Therefore, we consider the Poisson problem

$$\Delta \phi_{\text{mag}} = \nabla M \qquad \text{on } \Omega_1$$
$$\Delta \phi_{\text{mag}} = 0 \qquad \text{on } \Omega_2 \qquad (A.3)$$
$$\nabla \phi_{\text{mag}} = 0 \qquad \text{on } \partial \Omega_2$$

with Ω_1 a circle of radius 0.2, Ω_2 a circle of radius 1 and $M = \hat{y}$. This is the two-dimensional counter part of the homogeneously magnetized sphere on a finite domain. The weak formulation is given by

$$-\int_\Omega \langle \nabla \psi, \nabla \phi_{\text{mag}} - M \rangle dx = 0 \qquad \forall \psi \in H_0^1(\Omega) \qquad \text{on } \Omega_1$$

and $$-\int_\Omega \langle \nabla \psi, \nabla \phi_{\text{mag}} \rangle dx = 0 \qquad \forall \psi \in H_0^1(\Omega) \qquad \text{on } \Omega_2$$

For the solution process, a so-called *fem-structure* needs to be created, containing all the information on the model:

```
appl.mode.class='FlPDEW';         % Choice of application mode (weak)
appl.dim={'phi' 'phi_t'};         % Setting of dependent variables (phi)
appl.shape={'shlag(2,''phi'')'};  % Setting of test functions (2nd order
                                    Lagrange)
```

A.1 COMSOL Multiphysics^{TM}

```
appl.equ.weak={'-phix*test(phix)-phiy*test(phiy)'; ...
    '-(-phix-Mx)*test(phix)-(phiy-My)*test(phiy)'};
                            % Equations for different domain types
appl.equ.dweak='0';         % Time dependent contribution
appl.equ.ind=[1 2];         % Assigning equations to domains
appl.bnd.constr='-phi';     % Setting of boundary conditions (phi=0)
appl.bnd.ind=ones(1,8);     % Assigning conditions to boundaries
fem.appl{1}=appl;
fem.geom=geomcsg({circ2(1),circ2(0.2)});
                            % Definition of geometry
geomplot(fem);              % View geometry (Figure A.2(a))
fem=multiphysics(fem);      % Initializing PDE-system
fem.mesh=meshinit(fem);     % Initializing mesh
meshplot(fem);              % View mesh (Figure A.2(b))
fem.xmesh=meshextend(fem);  % Initializing of finite elements
fem.sol=femlin(fem);        % Solving the model
postplot(fem,'tridata','phi','arrowdata',{'phix' 'phiy'})
                            % View solution (Figure A.2(c))
```

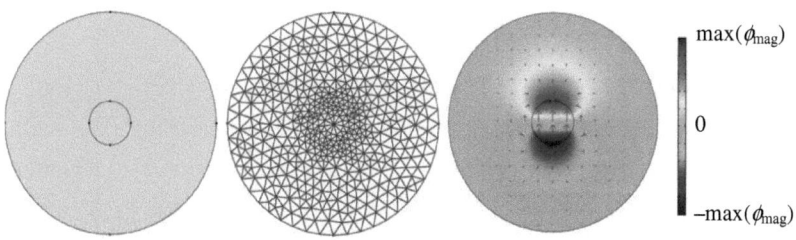

Figure A.1: Model creation by COMSOL Multiphysics^{TM}. (a) Model geometry, (b) finite element mesh, (c) solution of the problem. The colour code indicates the behaviour of ϕ, the arrows point in the direction of the gradient of ϕ.

Appendix

A.2 Magnetic point-particle under an external force

A small magnetic particle of radius R, mass m and magnetic moment $\boldsymbol{m}_{\text{part}}$ suspended in a liquid feels two different types of forces: a drag contribution due to the velocity difference between the velocity $\boldsymbol{v}$ of the particle and the velocity $\boldsymbol{u}$ of the liquid and external forces, summarized in $\boldsymbol{F}_{\text{ext}}$. Newton's second law applied on this system reads

$$m\frac{d\boldsymbol{v}}{dt} = \boldsymbol{F}_{\text{mag}} + \boldsymbol{F}_{\text{drag}} = \boldsymbol{F}_{\text{mag}} + 6\pi\eta R(\boldsymbol{u}-\boldsymbol{v}) \tag{A.4}$$

if we assume a creeping flow problem ($Re \ll 1$). A solution of the homogenenous system is given by

$$\frac{d\boldsymbol{v}}{dt} = -\frac{6\pi\eta R}{m}\boldsymbol{v} \quad\Rightarrow\quad \boldsymbol{v}(t) = \boldsymbol{C}\exp\left(-\frac{6\pi\eta R}{m}t\right)$$

with an integration constant $\boldsymbol{C}$. Varying the constant, we obtain for the inhomogeneous case

$$\frac{d\boldsymbol{C}}{dt} = \left(\frac{6\pi\eta R}{m}\boldsymbol{u} + \frac{\boldsymbol{F}}{m}\right)\exp\left(\frac{6\pi\eta R}{m}t\right) \tag{A.5}$$

Though the velocity profile $\boldsymbol{u}$ is time-independent in the creeping flow regime, an implicit time dependence is introduced by the particle position $\boldsymbol{r}(t)$. Therefore, we have $\boldsymbol{u} = \boldsymbol{u}(\boldsymbol{r}(t)) = \boldsymbol{u}(t)$ and (A.5) cannot be readily integrated in respect to time. Instead, it is

$$\boldsymbol{C}(t) = \boldsymbol{C}_0 + \int_{t_0}^{t}\frac{d\boldsymbol{C}(\tau)}{dt}d\tau = \boldsymbol{C}_0 + \left.\frac{\boldsymbol{F}_{\text{mag}}}{6\pi\eta R}\exp\left(\frac{6\pi\eta R}{m}t\right)\right|_{0}^{t} + \frac{6\pi\eta R}{m}\int_{0}^{t}\boldsymbol{u}(\tau)\exp\left(\frac{6\pi\eta R}{m}\tau\right)d\tau$$

$$= \boldsymbol{C}_0 + \frac{\boldsymbol{F}_{\text{mag}}}{6\pi\eta R}\left(\exp\left(\frac{6\pi\eta R}{m}t\right)-1\right)$$

$$+\boldsymbol{u}(t)\exp\left(-\frac{6\pi\eta R}{m}t\right) - \boldsymbol{u}(t_0) - \int_{0}^{t}\frac{d\boldsymbol{u}}{dt}\exp\left(\frac{6\pi\eta R}{m}\tau\right)d\tau$$

If we further assume a non-moving particle for $t=0$, we find $C_0 = 0$ and therefore obtain for the particle velocity v

$$v = \frac{F_{mag}}{6\pi\eta R}\left(1 - \exp\left(-\frac{6\pi\eta R}{m}t\right)\right)$$
$$+ u(t) - u(0)\exp\left(-\frac{6\pi\eta R}{m}t\right) - \int_0^t \frac{du}{dt}\exp\left(-\frac{6\pi\eta R}{m}\tau\right)d\tau \quad (A.6)$$

This is the general solution of (A.4). However, if we take a look on the values of different material parameters, we find $R = \mathcal{O}(10^{-8}\,\text{m})$ and $m = \mathcal{O}(10^{-20}\,\text{kg})$. Together with a viscosity of about $\eta = \mathcal{O}(10^{-3}\,\text{Pa s})$, we have $6\pi\eta R/m = \mathcal{O}(10^{10})$ and, therefore, the exponential function is close to the Kronecker δ-distribution

$$\exp\left(-\frac{6\pi\eta R}{m}t\right) \approx \delta(t) \quad \Rightarrow \quad v = u + \frac{F_{mag}}{6\pi\eta R} \quad \text{for } t > 0$$

Inertia effects may thus be omitted.

A.3 Derivation of the Brown equation

For the sake of completeness and to provide some reference material to everybody trying to extend the weak forms by another term not discussed in this thesis, we will explain how the individual energy contributions need to be varied in respect to the magnetization distribution. Similar to the "normal" derivative, the variational derivative is the linear change of a functional I in respect to a slight variation of the argument. In principle, this means we investigate the difference $I(f) - I(f + \delta f)$ with an arbitrary argument f and variation δf. If it possible to expand this expression in a power series of δf, we might write

$$I(f) - I(f + \delta f) = a_1\delta f + \frac{a_2}{2}(\delta f)^2 + \frac{a_3}{6}(\delta f)^3 + \dots \quad (A.7)$$

Appendix

The linear change is apparently given by the coefficient function a_1. Therefore, we try to calculate this function by expanding $I(f+\delta f)$ in powers of δf. According to chapter 3, the total energy under the constraint of $|\hat{m}|=1$ is given by

$$\tilde{E}[\hat{m},\phi_{mag},\lambda_m] = \int_{\Omega_{mag}} \left(A\left((\nabla \hat{m}_x)^2 + (\nabla \hat{m}_y)^2 + (\nabla \hat{m}_z)^2\right) + e_{ani} \right.$$
$$\left. + \frac{\mu_0}{2}\langle \nabla \phi_{mag}, M\rangle - \mu_0\langle M, H_{ext}\rangle \right) dr + \int_{\Omega_{mag}} \lambda_m(\hat{m}_x^2 + \hat{m}_y^2 + \hat{m}_z^2 - 1)\, dr$$

As the easiest case, we will begin our discussion with the Zeeman-energy. The effective functional may be written in the form

$$\tilde{E}_{Zeeman}[\hat{m},\lambda_m] = \int_{\Omega_{mag}} \left(-\mu_0\langle M, H_{ext}\rangle + \lambda_m(\hat{m}_x^2 + \hat{m}_y^2 + \hat{m}_z^2 - 1)\right) dr.$$

$$\tilde{E}_{Zeeman}[\hat{m}_x + \delta m_x, \hat{m}_y, \hat{m}_z, \lambda_m]$$
$$= \int_{\Omega_{mag}} \left(-\mu_0 M_S((\hat{m}_x + \delta m_x)H_{ext,x} + \hat{m}_y H_{ext,y} + \hat{m}_z H_{ext,z})\right) dr$$
$$+ \lambda_m \int_{\Omega_{mag}} ((\hat{m}_x + \delta m_x)^2 + \hat{m}_y^2 + \hat{m}_z^2 - 1)\, dr$$
$$= \int_{\Omega_{mag}} \left(-\mu_0\langle M, H_{ext}\rangle + \lambda_m(\hat{m}_x^2 + \hat{m}_y^2 + \hat{m}_z^2 - 1)\right) dr$$
$$+ \int_{\Omega_{mag}} \left(-\mu_0 M_S H_{ext,x}\delta m_x + 2\lambda_m \hat{m}_x \delta m_x\right) dr + \int_{\Omega_{mag}} \lambda_m(\delta m_x)^2\, dr$$
$$= \tilde{E}_{Zeeman}[\hat{m},\lambda_m] + \int_{\Omega_{mag}} \left(-\mu_0 M_S H_{ext,x} + 2\lambda_m \hat{m}_x\right)\delta m_x\, dr + \mathcal{O}((\delta m_x)^2)$$

According to definition (A.7), the variational derivative is given by the integrand of the second term. Similar results are obtained for the remaining magnetization components. Therefore, we may write

$$\frac{\delta E_{Zeeman}}{\delta \hat{m}} = -\mu_0 M_S H_{ext} + 2\lambda_m \hat{m} \tag{A.8}$$

A.3 Brown equations

The equilibrium state is given by the distribution which minimizes I. Again, similar to scalar functions, a necessary criterion for a local minimum is that the "first derivative" equals zero. Therefore, we have for $j, k \in \{x, y, z\}$

$$\frac{\delta E_{\text{Zeeman}}}{\delta \hat{m}_j} = -\mu_0 M_S H_{\text{ext},j} + 2\lambda_m \hat{m}_j \overset{!}{=} 0 \overset{!}{=} -\mu_0 M_S H_{\text{ext},k} + 2\lambda_m \hat{m}_k$$

$$\Rightarrow \quad \frac{\mu_0 M_S H_{\text{ext},j}}{2\lambda_m \hat{m}_j} = 1 = \frac{\mu_0 M_S H_{\text{ext},k}}{2\lambda_m \hat{m}_k} \quad \Leftrightarrow \quad M_S \hat{m}_k H_{\text{ext},j} = M_S \hat{m}_j H_{\text{ext},k}$$

$$\Leftrightarrow \quad M_S (\hat{m}_j H_{\text{ext},k} - \hat{m}_k H_{\text{ext},j}) = 0$$

The last equation may readily be written in the form $\varepsilon_{ijk} \hat{m}_j H_{\text{ext},k} = 0$ or

$$\hat{\boldsymbol{m}} \times \boldsymbol{H}_{\text{ext}} = \boldsymbol{0} \tag{A.9}$$

In a similar way, we may write for the exchange energy

$$\tilde{E}_A[\hat{\boldsymbol{m}}, \lambda_m] = \int_{\Omega_{\text{mag}}} \left(A\left((\nabla \hat{m}_x)^2 + (\nabla \hat{m}_y)^2 + (\nabla \hat{m}_z)^2\right) + \lambda_m (\hat{m}_x^2 + \hat{m}_y^2 + \hat{m}_z^2 - 1) \right) d\boldsymbol{r}$$

$$\Rightarrow \quad \tilde{E}_A[\hat{m}_x + \delta \hat{m}_x, \hat{m}_y, \hat{m}_z, \lambda_m] = \int_{\Omega_{\text{mag}}} \left(A\left((\nabla (\hat{m}_x + \delta m_x))^2 + (\nabla \hat{m}_y)^2 + (\nabla \hat{m}_z)^2\right) \right) d\boldsymbol{r}$$

$$+ \int_{\Omega_{\text{mag}}} \lambda_m ((\hat{m}_x + \delta m_x)^2 + \hat{m}_y^2 + \hat{m}_z^2) d\boldsymbol{r}$$

$$\Rightarrow \quad \tilde{E}_A[\hat{m}_x + \delta m_x, \hat{m}_y, \hat{m}_z, \lambda_m] = \tilde{E}_A[\hat{\boldsymbol{m}}, \lambda_m]$$

$$+ \int_{\Omega_{\text{mag}}} 2A \langle \nabla \hat{m}_x, \nabla \delta m_x \rangle d\boldsymbol{r}$$

$$+ \int_{\Omega_{\text{mag}}} 2\lambda_m \hat{m}_x \delta m_x \, d\boldsymbol{r} + \mathcal{O}((\delta m_x)^2)$$

$$\Rightarrow \quad \tilde{E}_A[\hat{m}_x + \delta m_x, \hat{m}_y, \hat{m}_z, \lambda_m] = \tilde{E}_A[\hat{\boldsymbol{m}}, \lambda_m]$$

$$- \int_{\Omega_{\text{mag}}} 2A \Delta \hat{m}_x \delta m_x \, d\boldsymbol{r} + \int_{\partial \Omega_{\text{mag}}} 2A \langle \nabla \hat{m}_x, \hat{\boldsymbol{n}} \rangle \delta m_x \, d\boldsymbol{r}$$

$$+ \int_{\Omega_{\text{mag}}} 2\lambda_m \hat{m}_x \delta m_x \, d\boldsymbol{r} + \mathcal{O}((\delta m_x)^2)$$

Appendix

Therefore, the integrals expanded over the magnetic volume and the surface of the magnetic material lead to domain contributions as well as a boundary condition which are given by

$$\frac{\delta E_A}{\delta \hat{m}} = -2A\Delta\hat{m} + 2\lambda_m \hat{m} \stackrel{!}{=} \mathbf{0} \quad \text{and} \quad \frac{\partial \hat{m}}{\partial \hat{n}} = \mathbf{0}, \tag{A.10}$$

respectively. Analogously to the case of Zeeman-energy, (A.10) may be rewritten in the form

$$-\hat{m} \times \frac{2A}{\mu_0 M_S} \Delta\hat{m} = \mathbf{0} \tag{A.11}$$

For the anisotropy energy, we will restrict our analysis to the uniaxial case only. Therefore, we may write

$$\tilde{E}_{ani}[\hat{m}, \lambda_m] = \int_{\Omega_{mag}} \left(K_1(1 - \langle \hat{m}, \hat{e} \rangle^2) + \lambda_m (\hat{m}_x^2 + \hat{m}_y^2 + \hat{m}_z^2 - 1) \right) dr$$

with $\hat{e}$ the easy axis. Similar to the cases considered above, we have

$$\tilde{E}_{ani}[\hat{m}_x + \delta m_x, \hat{m}_y, \hat{m}_z, \lambda_m] = \int_{\Omega_{mag}} K_1(1 - ((\hat{m}_x + \delta m_x)\hat{e}_x + \hat{m}_y \hat{e}_y + \hat{m}_z \hat{e}_z)^2) dr$$
$$+ \int_{\Omega_{mag}} \lambda_m((\hat{m}_x + \delta \hat{m}_x)^2 + \hat{m}_y^2 + \hat{m}_z^2 - 1) dr$$

$$\Rightarrow \quad \tilde{E}_{ani}[\hat{m}_x + \delta m_x, \hat{m}_y, \hat{m}_z, \lambda_m] = \int_{\Omega_{mag}} K_1(1 - \langle \hat{m}, \hat{e} \rangle^2) dr$$
$$+ \int_{\Omega_{mag}} 2K_1 \langle \hat{m}, \hat{e} \rangle \delta m_x \, dr + \mathcal{O}((\delta m_x)^2)$$
$$+ \int_{\Omega_{mag}} \lambda_m (\hat{m}_x^2 + \hat{m}_y^2 + \hat{m}_z^2 - 1) dr + \int_{\Omega_{mag}} 2\lambda_m \hat{m}_x \delta m_x \, dr + \mathcal{O}((\delta m_x)^2)$$

$$\Rightarrow \quad \tilde{E}_{ani}[\hat{m}_x + \delta m_x, \hat{m}_y, \hat{m}_z, \lambda_m] = \tilde{E}_{ani}[\hat{m}, \lambda_m] + \int_{\Omega_{mag}} 2K_1 \langle \hat{m}, \hat{e} \rangle \hat{e}_x \delta m_x \, dr$$
$$+ \int_{\Omega_{mag}} 2\lambda_m \hat{m}_x \delta m_x \, dr + \mathcal{O}((\delta m_x)^2)$$

and therefore

A.3 Brown equations

$$\frac{\delta E_{\text{ani}}}{\delta \hat{m}} = 2K_1 \langle \hat{m}, \hat{e} \rangle \hat{e} + 2\lambda_m \hat{m} \stackrel{!}{=} \mathbf{0} \quad \Rightarrow \quad \hat{m} \times \frac{2K_1}{\mu_0 M_S} \langle \hat{m}, \hat{e} \rangle \hat{e} = \mathbf{0}. \quad (A.12)$$

Finally, the demagnetization field is to be varied. The potential ϕ_{mag} needs to varied under the constraint of ϕ_{mag} satisfying the Poisson equation $\Delta \phi_{\text{mag}} = \nabla M$. A second Lagrange multiplier λ_φ is introduced in order to ensure this. The functional for the demagnetization energy may thus be written in the form

$$\tilde{E}[\hat{m}, \phi_{\text{mag}}, \lambda_m, \lambda_\varphi] = \frac{\mu_0}{2} \int_{\Omega_{\text{mag}}} (\nabla \phi_{\text{mag}})^2 + \lambda_\varphi (\Delta \phi - M_S \nabla \hat{m}) + \lambda_m (\hat{m}_x^2 + \hat{m}_y^2 + \hat{m}_z^2 - 1) \, dr$$

$$\Rightarrow \quad \tilde{E}[\hat{m}_x + \delta m_x, \hat{m}_y, \hat{m}_z, \phi_{\text{mag}}, \lambda_m, \lambda_\varphi]$$

$$= \frac{\mu_0}{2} \int_{\Omega_{\text{mag}}} (\nabla \phi_{\text{mag}})^2 + \lambda_\varphi \left(\Delta \phi - M_S \left(\nabla \hat{m} + \frac{\partial \delta \hat{m}_x}{\partial x} \right) \right) dr$$

$$+ \int_{\Omega_{\text{mag}}} \lambda_m (\hat{m}_x^2 + \hat{m}_y^2 + \hat{m}_z^2 - 1) \, dr + \int_{\Omega_{\text{mag}}} 2\lambda_m \hat{m}_x \delta m_x \, dr + \mathcal{O}((\delta m_x)^2)$$

$$\Rightarrow \quad \tilde{E}[\hat{m}_x + \delta m_x, \hat{m}_y, \hat{m}_z, \phi_{\text{mag}}, \lambda_m, \lambda_\varphi] = \tilde{E}[\hat{m}, \phi_{\text{mag}}, \lambda_m, \lambda_\varphi]$$

$$+ \int_{\Omega_{\text{mag}}} M_S \frac{\partial \lambda_\varphi}{\partial x} \delta \hat{m}_x \, dr - \int_{\partial \Omega_{\text{mag}}} M_S \lambda_\varphi \delta \hat{m}_x \, dr$$

$$+ \int_{\Omega_{\text{mag}}} 2\lambda_m \hat{m}_x \delta m_x \, dr + \mathcal{O}((\delta m_x)^2)$$

$$\Rightarrow \quad \frac{\delta E_{\text{demag}}}{\delta \hat{m}} = M_S \nabla \lambda_\varphi + 2\lambda_m \hat{m} \stackrel{!}{=} \mathbf{0} \quad (A.13)$$

Unfortunately, (A.14) cannot be readily reformulated in the form $\hat{m} \times \boldsymbol{H}_{\text{demag}} = \mathbf{0}$ since the demagnetization field introduces additional degrees of freedom. Therefore, the variation of ϕ_{mag} needs to be considered. Since this is similar to the cases above, we will only report that the resulting equation may be readily employed to obtain the corresponding formula. Combining all contribution, this finally leads to

$$\hat{m} \times \boldsymbol{H}_{\text{eff}} = \mathbf{0} \quad \text{with} \quad \boldsymbol{H}_{\text{eff}} = -\frac{2A}{\mu_0 M_S} \Delta \hat{m} + \frac{2K_1}{\mu_0 M_S} \langle \hat{m}, \hat{e} \rangle \hat{e} + \boldsymbol{H}_{\text{demag}} + \boldsymbol{H}_{\text{ext}}$$

Appendix

A.4 A short introduction to PADIMA

COMSOL Multiphysics™ provides a graphical user interface for the setup of the model. However, the systems discussed here, usually consist of several thousands of characters, a direct initialization of the problem is therefore very elaborate. In the framework of this thesis, the GUI-based plug-in PADIMA was developed to ease the setup and also automatically apply some algorithms for meshing, preconditioning etc. Since COMSOL also provides many predefined physics settings a further development of this tool by implementing further influences (external stress, coupling to spin currents, influence of temperature etc.) should be rather easy in the near future.

The graphical user interfaces aims to make the plug-in applicable to everybody not having worked with COMSOL Multiphysics™ before and allows an easy calculation of micromagnetic thin film systems under the influence of external (particle-) perturbations. Since I have written neither an instruction manual nor an easy to understand documentation of the code itself, this paragraph should give a short introduction for everybody who is interested in similar questions as investigated in this work. In general, all parameters (except the field values for the hysteresis analysis) should be given in SI-units, however, the units to be entered are also always given next to the corresponding text fields.

The interface is shown in Figure A.2. The setup of a model is done in three steps: **A particle properties** definition, **B layer properties** definition, and **C solving and postprocessing**. Figure A.2 shows the particle-panel. Geometrical properties of the

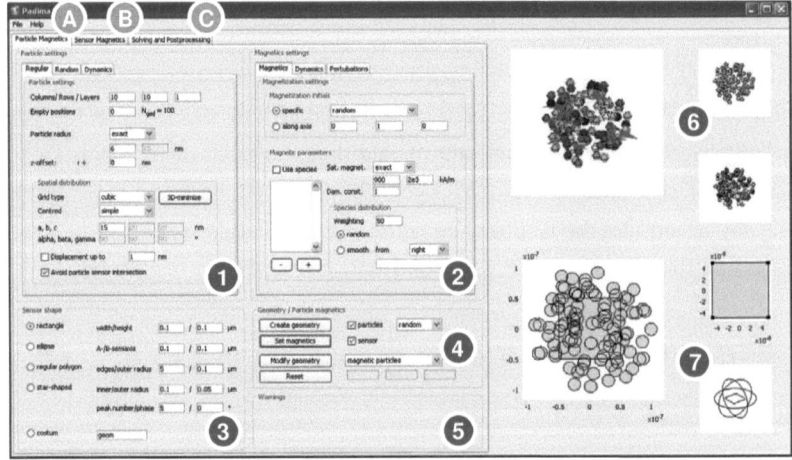

Figure A.2: Graphical user interface of COMSOL plug-in

A.4 PADIMA

particle assembly are defined in panel (1). Different ways of definition are possible: (a) **regular**, (b) **random**, and (c) **dynamic** (compare Figure A.3).

(a) regular: particles are arranged in a regular assembly of N_x **Columns**, N_y **Rows** and N_z **Layers** whereas N_h holes or **empty positions** are randomly distributed on the grid. For the definition of the **particle radius** different choices are possible:

- *exact*: a fixed value
- *between*: random values between the specified bounds
- *log-normal*: log-normal distribution with specified expectation value and standard deviation

As a default setting, the z-position of the assembly is chosen so that the "lowest" geometry point coincides with $z = 0$. This value may be changed to $z = $ *offset* by defining an additional **z-offset** value.

The **grid type** of the particle assembly may be chosen from **cubic-**, **tetragonal-**, **rhombic-**, **hexagonal-**, **trigonal-**, **monoclinic-**, and **triclinic-symmetries**. Additionally, the **centre**-type can be set to **simple-**, **base-**, **body-** or **face-centred**. The necessary geometry parameters (size lengths and angles of unit cells) can be specified in the text fields below. The **'3D-minimze'**-operation creates the cell of a minimal volume with the defined symmetry; in particular, it does not affect the angle settings. The perfect lattice may be disturbed by adding an additional **displacement** δ : Each particle position is shifted by a random value in $[0, \delta]$ into a random direction. If the **'Avoid particle sensor intersection'**-checkbox is enabled, all particles are situated at $z \geq 0$.

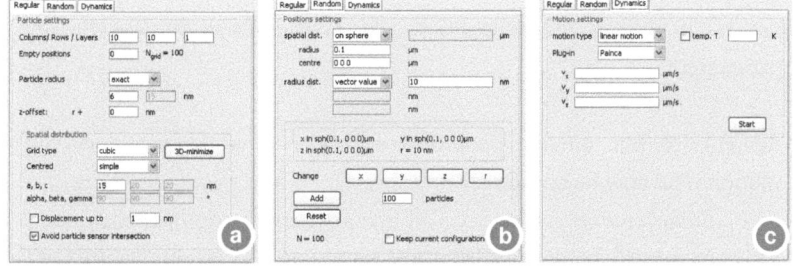

Figure A.3: Different possibilities for the particle setup: (a) regular, (b) random, and (c) dynamic.

151

Appendix

(b) random: particles are placed randomly. The choice for every component and the particle radius may be specified individually. For the components, different choices are possible:
- *vector value*: a single component is given by either a scalar (all particles have the same value) or a vector with a length that equals the number of particles.
- *between*: a single component is chosen randomly between the specified bounds.
- *on sphere*: all components are equally randomly distributed along a sphere of defined radius and centre. For the notation of the centre vector use either $[x\ y\ z]$ or simply $x\ y\ z$ with x, y, z the centre coordinates.

The radius options are given by:
- *vector value*: either a scalar (all particles have the same radius) or a vector with a length that equals the number of particles.
- *between*: random values between the specified bounds.
- *log-normal*: log-normal distribution with specified expectation value and standard deviation.

To specify either a certain component of the radius settings, lock the corresponding **Change**-button and **Add** the specified number of particles. If the **Add**-button is pressed again, additional particles are added to the configuration, the actual total number is shown below the **Reset**-button which deletes the particle definitions. The algorithms applied for the random creation generate non-overlapping particles. If the chosen settings do not allow for the specified number of particles (e.g. place 100 particles of 1 μm on 5 × 5 μm-square), the process is aborted and returns a smaller configuration that satisfies the input parameters.

(c) dynamic: define moving particles for (macro-)time-dependent systems. The motion type may be chosen from:
- *linear motion*: particles travel linearly between defined positions
- *in velocity field*: particles travel according to a given velocity field
- *from Plug-in*: particle motion is calculated via an external Plug-in (in the current version only the Plug-in PAINCA has been implemented which

A.4 PADIMA

may be employed for the calculation of the motion of particles in an external rotating magnetic field, compare also section 4.6.1)

Additional Brownian motion may be added via the **Temperature** T.

After the definition of the geometry settings, the particle **Magnetism** (2) needs to be defined. This is done in three steps (d) **magnetism**, (e) **dynamics**, and (f) **perturbations**.

(d) magnetism: The initial state, the **magnetization initials**, can be chosen from different possibilities borrowed from standard micromagnetic solutions:
- *random*: orientations are distributed equally randomly on the unit sphere surface
- *S-state*: orientations are ordered in an S-state
- *C-state*: orientations are ordered in a C-state
- *vortex-state*: orientations are ordered in a vortex-state
- *alternating*: adjacent particles have antiparallel alignment
- *current solution*: apply a previous solution as the new initial configuration

Alternatively, the **'along axis'**-setting aligns all moments in the direction specified. Axis length does not need to be normalized. Magnetic properties are defined in the **Magnetic parameters** panel. The **saturation magnetization** can be set to either an *exact* value or *between* two specified bounds. For time-dependent problems, the dimensionless damping constant α (compare section

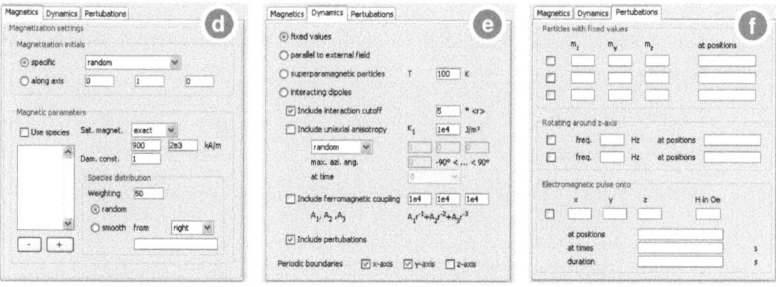

Figure A.4: Setup of the particle magnetism by defining (a) magnetism, (b) dynamics, and (c) perturbations.

Appendix

3.5) needs to be specified. The default value is set to $\alpha = 1$ to obtain fast convergence according to section 4.4. It is also possible to introduce different particle species. Therefore, set the material parameters and add this type via the '+'-button. A certain species may be removed again by marking it in the listbox and activating the '−'-button. The weighting of individual species is controlled via the **distribution**-panel. Either add particle with a certain **weighting** in a **random** way, or choose a **smooth** option:
- *right / left / top / bottom*: gradual decay from the chosen side to the opposite one
- *centre*: gradual decay from the centre to the outside
- *rows / columns*: position species along certain rows or columns only (only available for regular configurations)

(e) dynamics: define the dynamic behaviour of the magnetic particle moments. Four different options are available: **fixed values** (particles remain in their initial state), **parallel to external field** (particles will perfectly align to an external field), **superparamagnetic particles** (similar to before, but the saturation depends according to the Langevin-formalism from the **temperature** T), and **interacting dipoles** (particles interact dynamically according to the system of ODEs introduced in section 4.3). Considering interacting dipoles, additional options may be adjusted:
- *Interaction cutoff*: a measure of what particles interact with each other. In the default setting, it is set to five times the average particle radius.
- *Uniaxial anisotropy*: an additional uniaxial anisotropy with anisotropy constant K_1 is added. The direction may be set to *random* (along the unit sphere), *along axis* (set a specific direction for all particles), *normal to* (random in a specified plane), and *from solution* (use a solution for the new anisotropy settings).
- *Ferromagnetic coupling*: add additional coupling terms, leading to a parallel alignment of adjacent moments (in contrast to the dipole coupling), which may be compared to the exchange contribution for ferromagnetic materials.
- *Include perturbations*: add external perturbations (see (f))
- *Periodic boundaries*: impose periodic conditions in certain space directions

(f) perturbations: the perturbations-tab allows to add additional local influences to the system, e.g. a short but strong electromagnetic pulse along a small region and analyze how such a perturbation is damped or travels through the system. Three different methods may be applied: **'Particles with fixed values'** allows for a fixation of the magnetization direction of certain magnetic moments. The position vector contains the labels of the particles which can be found in the graphical output after the geometry is generated. **'Rotating around z-axis'** enforce moments to rotate around the z-axis with a given frequency. If the method **'Electromagnetic pulse onto'** is chosen, an external field is applied to certain particles of the given durations at given time points. All methods may be applied simultaneously, if interferences are found, the priority goes from top to bottom.

Before the geometry can be created the shape of an eventual sensor needs to be specified (3). Here different possibilities are given. **Rectangular** elements of given **width/height**, **ellipses** with length of **A-/B-semiaxis**, **regular polygons** with a specified **number of edges** and a certain **outer radius**, as well as **star-shaped geometries** which need to be specified by their **inner and outer radius**, their **peak number** and a **rotation angle**. If a two-dimensional geometry has been created in the local folder, it can be loaded to the GUI by the **custom** option. The name of the geometry variable simply needs to be entered in the from: NAME.

The geometry/magnetism creation panel (4) offers options to create and manipulate the system geometry. Operating the **'Create geometry'**-button creates the geometry, **particles** and **sensor**, if the corresponding checkboxes are marked. For the particle setup, it may be chosen between *random-* or *regular*-settings. Operating the **'Set magnetics'**-button writes the equations for the particles' magnetization. The **'Modify geometry'**-button allows to change several details of an existing geometry:
- *magnetic particles*: change particles only
- *sensor shape*: change sensor only
- *offset*: change particle positions in respect to the sensor
- *rescale*: rescale the geometry by a given factor
- *surrounding sphere*: change the radius of the surrounding sphere for the (layer) stray field calculation

Appendix

Press **Reset** to delete all geometry settings. The warning-panel (5) returns warnings if several input parameters are unreadable. Information on the particle geometry and the

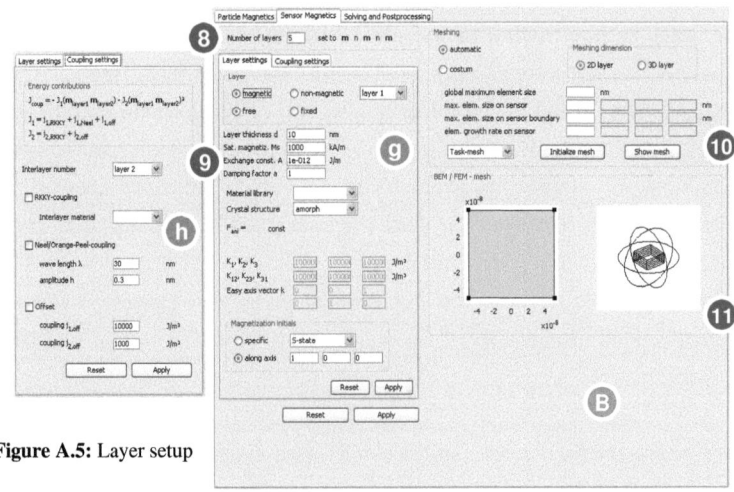

Figure A.5: Layer setup

magnetization settings can be found in the graphical output (6). The main plot shows the particle geometry, the small top plot the species settings and the lower one the magnetization configuration. The plots (7) show the relative ordering of particles and sensor, the sensor itself and the sensor together with the surrounding sphere.

The magnetic layers are configured in the tab **Sensor Magnetics** (B). A certain **Number of layers** (magnetic + non-magnetic) needs to be specified (8). The code behind the text box refers to the magnetism settings, **m** = magnetic, n = non-magnetic, properties may be configured below (9). Each individual layer may be adjusted in the **'Layer settings'**-tab by choosing it from the dropbox, the labelling counts from top to bottom. A layer may either be **magnetic** or **non-magnetic**. Every magnetic layer can additionally be **free** (only those contribute to the DOFs) or **fixed** creating a stray field in the surrounding area. Its orientation may be set via the initial configuration (see below). The layer is specified by its parameters **thickness**, **saturation magnetization**, **exchange constant**, and **damping constant**. If no material parameters are at hand, choose a material from the **material library**. Magnetocrystalline anisotropy may be added by the **'Crystal structure'**-settings. Different choices are available: *amorph, uniaxial (1st), uniaxial (2nd), cubic general, cubic [100], cubic [110], cubic [111], cubic/uniaxial, cubic/conical, orthorhombic* which require a certain amount of material parameters. Axis directions do not need to be entered in a normalized form. The **Magnetization initial**s may be set to either an *S-state*, a *C-state*, a *vortex state* or the *current solution* or the homogeneous configuration **along axis**. Changes are made permanent by operating the **'Apply'**-button or suspended by pushing the **'Reset'**-button.

Coupling energies are attributed to the non-magnetic interlayers (h). Choose the **Interlayer number** and apply different coupling phenomena:
- *RKKY-coupling*: the coupling strength is calculated according the thickness of the interlayer and the layer material chosen from the **'Interlayer material'**-library.
- *Néel/Orange-Peel-coupling*: the coupling strength is calculated according to section 5.1 evaluating the interlayer thickness as well as the **wave length** λ and the **amplitude** h.
- *Offset*: in order to integrate coupling without any physical reasons, bilinear and biquadratic coupling coefficients may be entered directly.

Changes are made permanent by operating the **'Apply'**-button or suspended by pushing the **'Reset'**-button.

Remark: If the layer number is reduced, the settings of suspended layers are still saved and will return if the layer number is increased again.

In order to retrieve the stack setup, operate the **Apply**-button in the **'Sensor magnetics'**-frame. The **Reset**-button restores the default values. The mesh menu (10) may be used to customize the finite element mesh. Use the **automatic**-option to obtain a mesh generated to ensure sufficient resolution. If the resulting number of DOFs is too high, it may be necessary to use a **custom**-mesh which allows for the individual definition of the mesh in each layer. The *Task-mesh*-option meshes two-dimensional and three-dimensional geometry at the same time. If they need to be meshed individually, use the *2D*- and *3D-mesh-option*. The meshing results are displayed in the plots (11).

The **Solving/Postprocessing**-tab provides different tasks and visualization options. Before the actual solving, the external magnetic field needs to be specified in the **External field settings** (12). The number of choices depends on the solver task, in principle though, the following settings are available:

Appendix

- *no external field*: set $H_{ext} = 0$
- *constant field*: set $H_{ext} = (H_x, H_y, H_z)^T$
- *alternating field*: the external field points parallel/antiparallel to a certain direction and changes linearly with a specified frequency

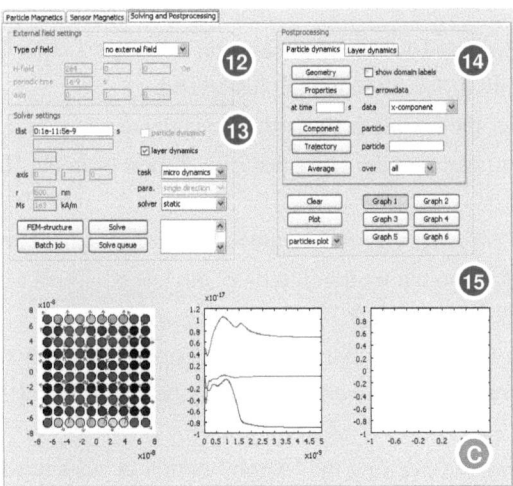

Figure A.6: Solving and post-processing tab.

- *rotating around z-axis*: an initial field vector of arbitrary direction rotates around the *z*-axis with a specified frequency

Highlight the **'particle dynamics'** and the **'layer dynamics'**-checkbox in the **Solver settings** (13) to include the respective component into the DOFs. If either of them has been specified but is not solved for, the behaviour is given by the fixed initial values. Different solver tasks may be employed

- *static*: find a static solution of the system
- *magnetic map*: calculate a system properties along a discrete set of parameters. In case of magnetic sensors, the option *MR-map* is available which calculates TMR-map (compare chapter 5) along a specified grid [**Xlist, Ylist, z**] with a specified probe particle. *High frequency* maps may be calculated for systems of interacting dipoles. The application mode calculates the response frequency spectra for a certain input frequency/frequency-list **flist**.
- *micro dynamics*: the time-dependent behaviour of a magnetic system for the time points **tlist** is analyzed solving Landau-Lifshitz Gilbert equation.

- *macro dynamics*: the time-dependent behaviour of a magnetic system for the time points **tlist** is analyzed solving the static equation of micromagnetics at each time point. This approach originates from the idea that all microprocesses are always finished on the macroscale. At each (macro-)time point the system is in a thermal equilibrium (due to the separation of time scale, compare section 2.1).
- *hysteresis*: calculates the equilibrium for a set of field values. The equilibrium state of the n-th step is always the initial guess of the $(n + 1)$-th step. Additionally to only consider a *single direction*, an *angle range* may be defined. The parametric solver is expanded over two parameters.
- *parametric*: consider the same system with different material parameters and solve for a parameter list **plist**.

The **Solve**-operation solves the specified task, whereas the fem-structure may be obtained by operating the **'FEM-structure'**-button. Additionally, one may **Batch jobs** and add them to a solving queue. The **'Solve queue'**-command solves collected tasks. As it turns out, homogeneous magnetization distributions are not consistent initial values for the systems. In order to initialize the solving process, a preconditioning method as shown in Figure A.7 is employed. For the definition of the Φ-functions, see section 5.1. The obtained consistent values may afterwards be employed for calculations. If a stationary solution needs to be found, a second preconditioning step is applied. The model is solved by time-dependent solvers for several steps in order to obtain a good initial guess.

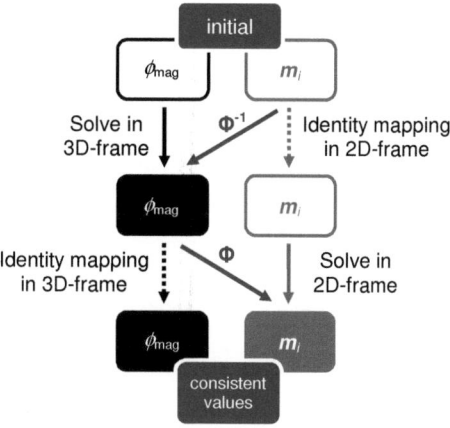

Figure A.7: Schematic representation of the solver sequence to find consistent initial values. The potential ϕ_{mag} and the magnetization distributions are treated as linearly coupled which is *no* solution but leads to a good initial guess for all DOFs.

The postprocessing-panel (14) allows for a number of different visualization techniques which are all displayed in the plots (15).

Appendix

References

Afshar R, Moser Y, Lehnert T, Gijs MAM, **2009**: *Magnetc particle dosing and size separation in a microfluidic channel*, Sens. Act. B: Chem., DOI: 10.1016/j.snb.2009.08.044

Albon C, **2009**: *High integration of tunneling magnetoresistive sensors for sensitive particle detection*, PhD thesis, Bielefeld University, Bielefeld

Amann H, Escher J, **2001**: *Analysis III*, 1st ed., Birkhäuser Verlag Basel, Boston, Berlin

Ascher UM, Petzold LR, **1998**: *Computer methods for ordinary differential equations and differential algebraic equations*, SIAM, Philadelphia

Auge A, Weddemann A, Wittbracht F, Hütten A, **2009**: *Magnetic ratchet for biotechnological applications*, Appl. Phys. Lett. *94* (18), 183507

Azevedo A, Oliveira AB, de Aguiar FM, Rezende SM, **2000**: *Extrinsic contribution to spin-wave damping and renormalization in thin $Ni_{50}Fe_{50}$ films*, Phys. Rev. B. *62*, 5331

Baibich MN, Broto JM, Fert A, Nguyen F, Petroff F, Etienne P, Creuzet G, Friederich A, Chazelas J, **1988**: *Giant Magnetoresistance of (001) Fe / (001) Cr Superlattices*, Phys. Rev. Lett. *61*, 2472

Batchelor GK, **1970**: *An introduction to fluid dynamics*, 2nd ed. Cambridge University Press, Cambridge

Berger L, **1996**: *Emission of spin waves by a magnetic multilayer traversed by a current*, Phys. Rev. B *54*, 9353

Beyn WJ, Thümmler V, **2004**: *Freezing solutions of equivariant evolution equations*, SIAM J. Appl. Dyn. Syst.

Beyn WJ, Selle S, Thümmler V, **2008**: *Freezing multipulses and multifronts*, SIAM J. Appl. Dyn. Syst.

Beyn WJ, Thümmler V, **2009**: *Dynamics of Patterns in Nonlinear Equivariant PDEs*, GAMM-Mitt. *32* (1), 7-25

Blattert C, **2005**: *Mikrofluidisches Trennverfahren für die chipintegrierte Blutdiagnostik*, PhD thesis, Albert-Ludwigs-Universität, Freiburg

Blizer C, Devolder T, Kim JV, Counil G, Chappert C, Cardoso S, Freitas PP, **2006**: *Study of the dynamic magnetic properties of soft CoFeB films*, J. Appl. Phys. *100*, 053903

Boffi D, Gastaldi L, **2004**: *Stability and geometric conversation laws for ALE formulations*. Comput. Meth. Appl. Mech. Engrg. *193*, 4717-4739

Bradley CJ, Cracknell AP, **1972**: *The mathematical theory of symmetry in solids*, Clarendon Press, Oxford

Brzeska M, Panhorst M, Kamp PB, Schotter J, Reiss G, Pühler A, Becker A, Brückl H, **2004**: *Detection and manipulation of biomolecules by magnetic carriers*, J. Biotech. *112*, 25-33

Brückl H, Panhorst M, Schotter J, Kamp PB, Becker A, **2005**: *Magnetic particles as markers and carriers of biomolecules*, IEE Proc. Nanobiotechnol. *152* (1), 41-46

Bruus H, **2008**: *Theoretical microfluidics*, Oxford University press Inc., New York

Butcher JC, **1987**: *The numerical analysis of ordinary differential equations—Runge-Kutta and general linear methods*, Wiley, Chichester

Carlo DD, Irimia D, Tompkins RG, Toner M, **2007**: *Continuous inertial focusing, ordering and separation of particles in microchannels*, Proc. Natl. Acad. Sci. USA *104*, 18892-18897

Carlo DD, **2009**: *Inertial microfluidics*, Lab Chip *9*, 3028-2046

References

Chang YC, Hou TY, Merriman B, Osher S, **1996**: *A level set formulation of eulerian interface capturing methods for incompressible fluid flows*, J. Comput. Phys. *124* (2), 449-464

Chen M, Kim J, Liu JP, Fan H, Sun S, **2006**: *Synthesis of FePt Nanocubes and their Oriented Self-Assembly*, J. Am. Chem. Soc. *128*, 7132-7133

Chernavskii PA, Peskov NV, Mugtasimov AV, Lunin VV, **2007**, Rus. J. Phys. Chem. B *1*, 394-411

Ciarlet PG, **1978**: *The finite element method for elliptic problems*, Amsterdam, New York, Oxford: North-Holland

Clay Mathematic Institute, **2009**, http://www.claymath.org/millennium/

COMSOL Multiphysics, **2005**: *COMSOL Multiphysics^{TM} Users' guide*, version 3.2, FEMLAB GmbH

Coroiu I, Cristea V, **2005**: *Proton NMR relaxivity of blood samples in the presence of iron, gadolinium and dysprosium compounds*, J. Magn. Magn. Mat. *293*, 520-525

Coulson JM, Richardson JF, **2003**: *Chemical Engineering, Vol. 2: Particle Technology and Separation Processes*, Butterworth-Heinemann, Oxford

DASPK: P.N. Brown, A.C. Hindmarsh, L.R. Petzold, **1994**: *Using Krylow methods in the solution of large-scale differential-algebraic systems*, SIAM J. Sci. Comput. *15*, 1467-1488

DASSL: K.E. Brenen, S.L. Campbell, L.R. Petzold, **1996**: *Numerical Solution of Initial-Value Problems in Differential-Algebraic Equations*, 2nd, Elsevier, New York

Davis TA, Duff IS, **1997**: *An unsymmetric-pattern multifrontal method for sparce LU factorization*, SIAM J. Matrix Anal. Applic. *18* (1), 140-158

Davis TA, Duff IS, **1999**: *A combined unifrontal/multifrontal method for unsymmetric sparse matrices*, ACM Trans. Math. Softw. *25* (1), 1-19

Deng T, Whitesides GM, Radhakrishnan, Zabow G, Prentiss M, **2001**: *Manipulation of magnetic microbeads in suspension using micromagnetic systems fabricated with soft lithography*, Appl. Phys. Lett. *78* (12), 1775-1777

Dittrich PS, Schwille P, **2003**: *An Integrated Microfluidic System for Reaction, High-Sensitivity Detection, and Sorting of Fluorescent Cells and Particles*, Anal. Chem. *75*, 5767-5774

Dobson J, **2006**: *Magnetic nanoparticles for drug delivery*, Drug Dev. Res. *67* (1), 55-60

Döring W, **1966**: *Mikromagnetismus*, in: *Handbuch der Physik*, Vol. 18, 2nd ed., Springer, Berlin, Heidelberg, New York, 341–437

Edelstein RL, Tamanaha CR, Sheehan PE, Miller MM, Baselt DR, Whitman LJ, Coltom RJ, **2000**: *The BARC biosensor applied to the detection of biological warfare agents*, Biosens. Bioelec. *14*, 805-813

Einstein A, **1905**: *On the motion of small particles suspended in liquids at rest required by the molecular kinetic theory of heat*, Ann. Phys. *17*, 549

Ennen I, Höink V, Weddemann A, Hütten A, Schmalhorst J, Reiss G, Waltenberg C, Jutzi P, Weis T, Engel D, Ehresmann A, **2007**: *Manipulation of magnetic nanoparticles by the strayfield of magnetically patterned ferromagnetic layers*, J. Appl. Phys. *102* (1), 013910

Ennen I, **2008**: *Magnetische Nanopartikel als Bausteine für granulare Systeme: Mikrostrukture, Magnetismus und Transporteigenschaften*, PhD thesis, Bielefeld University, Bielefeld

Faucheux LP, Libchaber AJ, **1994**: *Confined Brownian motion*, Phys. Rev. E *49* (6), 5158-5164

Ferreira HA, Graham DL, Freitas PP, **2003**: *Biodetection using magnetically labelled biomolecules and arrays of spin valve sensors*, J. Appl. Phys. *93*, 7281

Fonnum G, Johansson C, Molteberg A, Mørup S, Aksnes E, **2005**: *Characterisation of Dynabeads® by magnetization measurements and Mössbauer spectroscopy*, J. Magn. Magn. Mat. *293*, 41-47

Forster O, **1999**: *Analysis 3 – Integralrechnung im $\mathbb{R}^n$ mit Anwendungen*, 3rd ed., Vieweg, Braunschweig/Wiesbaden

Franca L, Frey S, Hughes T, **1992**: *Stabilized finite element methods: I. Application to the advective diffusive model*, Comput. Meth. Appl. Mech. Engrg. *95*, 253-276

Gao F, Pan BF, Zheng WM, Ao LM, Gu HC, **2005**: *Study of strepdavidin coated onto PAMAM dendrimer modified magnetite nanoparticles*, J. Magn. Magn. Mat. *293*, 48-54

Gijs MAM, **2004**: *Magnetic bead handling on-chip: new opportunities for analytical applications*, Microfluid. Nanofluid. *1*, 22-40

Gilbert TL, **1955**: *A Lagrangian formulation of the gyromagnetic equation of the magnetization field*, Phys. Rev. *100*, 1243

Golla JM, Gupta PK, Hung CT, Perrier DG, **2006**: *Evaluation of drug delivery following the administration of magnetic albumin microspheres containing adriamycin to the rat*, J. Pharm. Sci. *78* (3), 190-194

Gonis A, Butler WH, **2000**: *Multiple Scattering in Solids*, Springer, New York

Graham DL, Ferreira HA, Freitas PP, Cabral JMS, **2003**: *High sensitivity detection of molecular recognition using magnetically labelled biomolecules and magnetoresistive sensors*, Biosens. Bioelec. *18*, 483-488

Graham DL, Ferreira HA, Freitas PP, **2004**: *Magnetoresistiv-based biosensors and biochips*, Trends in Biotech. *22* (9), 455-462

Green NG, Morgan H, **1998**: *Separation of submicrometre particles using a combination of dielectric and magnetohydrodynamic forces*, J. Phys D: Appl. Phys. *31*, L25-L30

Green NG, Ramos A, Gonzáles A, Morgan H, Castellanos A, **2000**: *Fluid flow induced by nonuniform ac electric fields in electrolytes in microelectrodes*, Phys. Rev. E *61* (4), 4011-4018

Green NG, Ramos A, Gonzáles, Castellanos A, Morgan H, **2001**: *Electrothermally induced fluid flow on microelectrodes*, J. Electrostat. *53*, 71-87

Grünberg P, Schreiber R, Pang Y, Brodsky MD, Sowers H, **1986**: *Layered Magnetic Structures:Evidence for Antiferromagnetic Coupling of Fe Layers across Cr Interlayers*, Phys. Rev. Lett. *57*, 2442

Guillard H, Farhat C, **2000**: *On the significance of the geometrical conservation laws for flow computations on moving meshes*, Comput. Meth. Appl. Mech. Engrg. *190*, 1467-1482

Gupta PK, Hung CT, **1989**: *Magnetically controlled targeted micro-carrier systems*, Life-Sci. *44* (3),

Gupta PK, Hung CT, **1990**: *Trageted delivery of low dose doxorubicin administered via magnetic albumin microspheres in rats*, J. Microenc. *7* (1), 85-94

Guslienko KY, Novosad V, Otani Y, Shima H, Fukamichi, **2001**, *Field evolution of magnetic vortex state in ferromagnetic disks*, Appl. Phys. Lett. *78* (24), 3848-3850

Guslienko KY, Han XF, Keavney DJ, Divan R, Bader SD, **2006**: *Magnetic Vortex Core Dynamics in Cylindrical Ferromagnetic Dots*, Phys. Rev. Lett. *96*, 067205

Guslienko KY, Slavin AN, Tiberkevich V, Kim SK, **2008**: *Dynamic Origin of Azimuthal Modes Splitting in Vortex-State Magnetic Dots*, Phys. Rev. Lett. *101* (24), 247203

References

Hackbusch W, **1996**: *Gewöhnliche Differentialgleichungen*, Teubner, Stuttgart

Häfeli UO, Lobedann, Steingroewer, Moore LR, Riffle J, **2005**: *Optical method for measurement of magnetophoretic mobility of individual magnetic microspheres in defined magnetic field*, J. Magn. Magn. Mat. *293*, 224-239

Hanke-Bourgeois M, **2006**: *Grundlagen der Numerischen Mathematik und des Wissenschaftlichen Rechnens*, 2nd ed., Teubner, Stuttgart, Leipzig, Wiesbaden

Hedwig P, **2009**: *Dynamische Messung magnetischer Beads mit Hilfe eines TMR-Sensor-Arrays*, Diploma thesis, Bielefeld University, Bielefeld

Hrennikoff A, **1941**: *Solution of the Problems of Elasticity*, ASME J. Appl. Mech. *8*, A619-A715

Hubert A., Schäfer R, **2000**: *Magnetic domains: the analysis of magnetic microstructures*, Springer, Berlin

Hundsdorfer W, Verwer J, **2003**: *Numerical Solution of Time-Dependent Advection-Diffusion-Reaction-Equations*, Springer, Berlin, Heidelberg

Hütten A, Sudfeld D, Ennen I, Reiss G, Hachmann W, Heinzmann U, Wojczykowski K, Jutzi P, Saikaly W, Thomas G, **2004**: *New magnetic nanoparticles for biotechnology*, J. Biotech. *112*, 47-63

Hütten A, Sudfeld D, Ennen I, Reiss G, Wojczykowski K, Jutzi P, **2005**: *Ferromagnetic FeCo nanoparticles for biotechnology*, J. Magn. Magn. Mat. *293*, 93-101

ImageJ, **2009**, http://rsbweb.nih.gov/ij/

I-Sys, **2008**, http:// www.i-sys.de/

Jackson JD, **1975**: *Classical electrodynamics*, 2nd ed., Wiley, New York

Jahn A, Vreeland WN, Gaitan M, Locascio LE, **2004**: *Controlled vesicle self-assembly in microfluidic channels with hydrodynamic focusing*, J. Am. Chem. Soc. *126*, 2674-2675

Jiang Z, Llandro J, Mitrelias T, Bland JAC, **2006**: *An integrated microfluidic cell for detection, manipulation and sorting of single micron-sized magnetic beads*, J. Appl. Phys. *99*, 08S105

Jo BH, Van Lerberghe LM, Motsegood KM, Beebe DJ, **2000**: *Three-dimensional micro-channel fabrication in polydimethylsiloxane (PDMS) elstomer*, J. Microelectromech. Syst. *9*, 76-81

Jordan A, Wust P, Fähling H, John W, Hinz A, Felix R, **2009**: *Inductive heating of ferromagnetic particles and magnetic fluids: Physical evaluation of their potential for hyperthermia*, Inter. J. Hyper. *25* (7), 499-511

Julière M, **1975**: *Tunneling between ferromagnetic films*, Phys. Lett. *54*A, 225-226

Kamio E, Ono T, Yoshizawa H, **2009**: *Design of a new static micromixer having simple structure and excellent mixing performance*, Lap Chip *9*, 1809-1812

Kogan S, **1998**: *Electronic noise and fluctuations in solids*, 1st ed. Cambridge University Press Inc., Cambridge

Kreis HO, Lorenz J, **1989**: *Initial-boundary value problems and the Navier-Stokes equations*, Academic Press, London

Krishanu N, Chaudhuri S, Ganguly R, Puri IK, **2008**: *Analytical model for the magnetophoretic capture of magnetic microspheres in microfluidic devices*, J. Magn. Magn. Mat. *320*, 1398-1405

Kronmüller H, Hertel R, **2000**: *Computational micromagnetism of magnetic structures and magnetisation processes in small particles*, J. Magn. Magn. Mat. *215-216*, 11-17

Kronmüller H, **2007**: *Handbook of Magnetism and Advanced Magnetic Materials*, Vol. 2, John Wiley & Sons

Labrune M, Miltat J, **1995**: *Wall structures in ferro/ferromagnetic exchange-coupled bilayers: A numerical micromagnetic approach*, J. Magn. Magn. Mat. *151*, 231-245

Lacharme F, Vandevyver C, Gijs MAM, **2008**: *Full on-chip nanoliter immunoassay by geometrical magnetic trapping of nanoparticle chains*, Anal. Chem. *80* (8), 2905-2910

Lacharme F, Vandevyver, Gijs MAM, **2009**: *Magnetic beads retention device for sandwich immunoassay: comparison of off-chip and on-chip antibody incubation*, Microfluid. Nanofluid. *7* (4), DOI: 10.1007/s10404-009-0424-7

Landau LD, Lifshitz E, **1935**: *On the theory of the dispersion of magnetic permeability in ferromagnetic bodies*. Phys. Z. Sowjetunion *8*, 153-169

Landau LD, Lifshitz E, **1991**: *Lehrbuch der theoretischen Physik IV – Hydrodynamik*, 5th ed., Akademie Verlag GmbH, Berlin

Langer R, **1990**: *New methods of drug delivery*, Science *249*, 1527-1533

Langer R, **1998**: *Drug delivery and targeting*, Nature *392*, 5-10

Laroze D, Vargas P, Cortes C, Gutierrez G, **2009**: *Dynamic of two interacting dipoles*, J. Magn. Magn. Mat. *320*, 1440-1448

Larsson S, Thomée V, **2005**: *Partial Differential Equations with Numerical methods*, 2nd ed., Springer, Berlin, Heidelberg, 25-32

Lee CS, Lee H, Westervelt RM, **2001**: *Microelectromagnets for the control of magnetic nanoparticles*, Appl. Phys. Lett. *79* (20), 3308-3310

Lee SH, Kang HJ, Choi B, **2009**: *A study on the novel micromixer with chaotic flows*, Microsys. Tech. *15* (2), 269-277

Lehndorff R, Buchmeier M, Burgler DE, Kakay A, Hertel R, Schneider CM, **2007**: *Asymmetric spin-transfer torque in single-crystalline Fe/Ag/Fe nanopillars*, Phys. Rev. B *76* (21), 214420

Lehndorff R, Burgler DE, Kakay A, Hertel R, Schneider CM, **2008**: *Spin-transfer induced dynamic modes in single-crystalline Fe-Ag-Fe nanopillars*, IEEE Trans. Magn. *44* (7), 1951-1956

Lehndorff R, Burgler DE, Gliga S, Hertel R, Grünberg P, Schneider CM, **2009**: *Magnetization dynamics in spin torque nano-oscillators: Vortex state versus uniform state*, Phys. Rev. B *80* (5), 054412

Lesoinne M, Farhat C, **1996**: *Geometric conservation laws for flow problems and moving boundaries and deformable meshes and their impact on aerolastic computations*, Comput. Meth. Appl. Mech. Engrg. *134*, 71-90

Lin C, Lee G, Chang B, Chang GL, **2002**: *A new fabrication process for ultra-thick microfluidic utilizing SU-8 photoresist*, J. Micromech. Microeng. *12* (5), 590

Liu X, Brenner KH, Wilzbach M, Schwarz M, Fernholz T, Schmiedmayer J, **2005**: *Fabrication of alignment structures for a fiber resonator by use of deep-ultraviolet lithography*, Appl. Opt. *44*, 6857-6860

Liu Y, Jin W, Wang Z, **2006**: *Micromagnetic simulation for detection of a single magnetic microbead or nanobead by spin-valve sensors*, J. Appl. Phys. *99*, 08G102

Long M, Sprague MA, Grimes AA, Rich BD, Khine M, **2009**: *A simple three-dimensional vortex micromixer*, Appl. Phys. Lett. *94*, 133501

Loureiro J, Ferreira R, Cardoso S, Freitas PP, Germano J, Fermon C, Arrias G, Pannetier Lecoeur M, Rivadulla F, Rivas J, **2009**: *Toward a magnetoresistive chip cytometer: Integrated detec-*

References

tion of magnetic beads flowing at cm/s velocities in microfluidic channels, Appl. Phys. Lett. *95*, 034104

Macaroff PP, Oliveira DM, Ribeiro KF, Lacava ZGM, Lima ECD, Morais PC, Tedesco AC, **2005**: *Studies of cell toxicity of complexes of magnetic fluids and biological macromolecules*, J. Magn. Magn. Mat. *293*, 293-297

Magin RL, Bacic G, Niesman MR, Alameda JC, Wright SM, Swartz HM, **1991**: *Dextran magnetite as a liver contrast agent*, Magn. Reson. Med. *20* (1), 1-16

Mahmoudi M, Simchi A, Milani AS, Stroeve P, **2009**a: *Cell toxicity of superparamagnetic iron oxide nanoparticles*, J. Coll. Interf. Sci. *336* (2), 510-518

Mahmoudi M, Shokrgozar, Simchi A, Imani M, Milani AS, Stroeve P, Vali H, Häfeli UO, Bonakdar S, **2009**: *Multiphysics Flow Modeling and in Vitro Toxicity of Iron Oxide Nanoparticles Coated with Poly(vinyl alcohol)*, J. Phys. Chem. C *113*, 2322-2331

Masud A, **2006**: *Effects of mesh motion on the stability and convergence of ALE based formulations for moving boundary flows*. Comput. Mech. *38*, 430-439

Megens M, Prins M, **2005**: *Magnetic biochips: a new option for sensitive diagnostics*, J. Magn. Magn. Mat. *293*, 702-708

Meyners D, **2006**: *Herstellung und Charakterisierung von Logikarrays mit ultrakleinen magnetischen Tunnelelementen*, PhD thesis, Bielefeld University, Bielefeld

MicroCoat, **2008**, http://www.microcoat.de/

Mikkelsen C, Bruus H, **2005**a: *Microfluidic capturing-dynamics of paramagnetic bead suspensions*, Lab Chip *5*, 1293-1297

Mikkelsen C, Hansen MF, Bruus H, **2005**b: *Theoretical comparison of magnetic and hydrodynamic interactions between magnetically tagged particles in microfluidic systems*, J. Magn. Magn. Mat. *293*, 578-583

Mirowski E, Moreland J, Russek S, Donahue M, Hsieh K, **2007**: *Manipulation of magnetic particles by patterned arrays of magnetic spin-valve traps*, J. Magn. Magn. Mat. *311*, 401-404

Möller W, Takenaka, Buske N, Felten K, Heyder J, **2005**: *Relaxation of ferromagnetic nanoparticles in macrophages: In vitro and in vivo studies*, J. Magn. Magn. Mat. *293*, 245-251

Mornet S, Portier J, Duguet J, **2005**: *A method for synthesis and functionalization of ultrasmall supereparamagnetic covalent carriers based on maghemite and dextran*, J. Magn. Magn. Mat. *293*, 127-134

Moser A, Takano K, Margulies DT, Albrecht M, Sonobe Y, Ikeda Y, Sun S, Fullerton EE, **2002**: *Magnetic recording: advancing into the future*, J. Phys. D: Appl. Phys. *35*, R157-R167

MTrackJ, **2009**, http://www.imagescience.org/meijering/software/mtrackj/

Nader S, **2009**: *Hyperthermia and Cancer Treatment*, Heat Trans. Eng. *30* (12), 915-917

Néel L, **1962**, Acad. Sci. *255*, 1545; 1676

Nesliturk A, Harari I, **2003**: *The nearly-optimal Petrov-Galerkin method for convection-diffusion problems*, Comput. Meth. Appl. Mech. Engrg. *192*, 2501-2519

Niu X, Liu L, Wen W, Sheng P, **2006**: *Hybrid Approach of High-Frequency Microfluidic Mixing*, Phys. Rev. Lett. *97*, 044501

Nobile F, **2001**: *Numerical approximation of fluid structure interactions problems with applications in haemodynamics*, PhD thesis, Ecole polytechnique fédérale de Lausanne, Lausanne

Nochetto RH, Siebert KG, Veeser A, **2009**: *Theory of adaptive finite element mehtods: An introduction*, Springer, Berlin, Heidelberg, 409-542

Oden JT, **1976**: *An introduction to the mathematical theory of finite elements*, Wiley, New York

Osher S, Ronald F, **2003**: *Level Set Methods and Dynamic Implicit Surfaces*, Applied Mathematical Sciences *153*

Pamme N, Koyama R, Manz A, **2003**: *Counting and sizing of particles and particle agglomerations in a microfluidic device using laser light scattering: application to a particle-enhanced immunoassay*, Lab Chip *3*, 187-192

Pamme N, Manz A, **2004**: *On-Chip Free Flow Magnetophoresis: Continuous Flow Separation of Magnetic Particles and Agglomerates*, Anal. Chem. *76* (24), 7250-7526

Pamme N, Eijkel J, Manz A, **2006a**: *On-chip free flow magnetophoresis: Separation and detection of mixtures of magnetic particles in continuous flow*, J. Magn. Magn. Mat. *307* (2), 237-244

Pamme N, **2006b**: *Magnetism and microfluidics*, Lab Chip *6*, 24-38

Pamme N, Wilhelm C, **2006c**: *Continuous sorting of magnetic cells via on-chip free-flow magnetophoresis*, Lab Chip *6*, 974-980

Pamme N, **2007**: *Continuous flow separations in microfluidic devices*, Lab Chip *7*, 1644-1659

Pekas N, Granger M, Tondra, Popple A, Porter MD, **2005**: *Magnetic particle diverter in an integrated microfluidic format*, J. Magn. Magn. Mat. *293*, 584-588

Peyman SA, Iles A, Pamme N, **2009**: *Mobile magnetic particles as solid-support for rapid surface-based bioanalysis in continuous flow*, Lab Chip *9*, 3110-3117

Ravula SK, Branch DW, James CD, Townsend RJ, Hill M, Kaduchak G, Ward M, Brener I, **2008**: *A microfluidic system combining acoustic and dielectrophoretic particle preconcentration and focusing*, Sens. Act. B *130*, 645-652

Reiner, **2009**, http://www.reiner.de

Roch A, Gossuin Y, Muller RN, Gillis P, **2005**: *Superparamagnetic colloid suspensions: Water magnetic relaxation and clustering*, J. Magn. Magn. Mat. *293*, 532-539

Russ S, Bunde A, **2006**: *Monte Carlo simulations of frozen metastable states in ordered systems of ultrafine magnetic particles*, Phys. Rev. B. *74*, 064426

Salloum M, Ma RH, Weeks D, Zhu L, **2008**: *Controlling nanoparticle delivery in magnetic nanoparticle hyperthermia for cancer treatment: Experimental study in agarose gel*, Inter. J. Hyper. *24* (4), 337-345

Sawetzki T, Rahmouni S, Bechinger C, **2008**: ... D.W.M. Marr, Proc. Nat. Acad. Sci. *105*, 20141

Schaller V, Wahnströhm G, Sanz-Velasco A, Gustafsson S, Olsson E, Enoksson P, Johansson C, **2006**: *Effective magnetic moment of magnetic multicore nanoparticles*, Phys. Rev. B *80* (9), 092406

Schaller V, Wahnströhm G, Sanz-Velasco A, Enoksson P, Johansson C, **2009**: *Monte Carlo simulation of magnetic multi-core nanoparticles*, J. Magn. Magn. Mat.*321* (10), 1400-1403

Schepper W, Schotter J, Brückl H, Reiss G, **2004**: *Analysing a magnetic molecule detection system–computer simulation*, J. Biotech. *112*, 35-46

Schotter J, Kamp PB, Becker A, Pühler A, Reiss G, Brückl H, **2003**: *Comparison of a prototype magnetoresitive biosensor to standard fluorescent DNA detection*, Biosens. Bioelec.*19* (10), 1149-1156

Schultz MD, Calvin S, Fatouros PP, Morrison SA, Carpenter EE, **2007**: *Enhanced ferrite nanoparticles as MRI contrast agents*, J. Magn. Magn. Mat. *311*, 464-468

Sensitec, **2008**, http://www.sensitec.com

References

Shen W, Schrag BD, Carter MJ, Xie J, Xu C, Sun S, Xiao G, **2008**a: *Detection of DNA labeled with magnetic nanoparticles using MgO-based magnetic tunnel junction sensors*, J. Appl. Phys. *103*, 07A306

Shen W, Schrag BD, Carter MJ, Xiao G, **2008**b: *Quantitative detection of DNA labeled with magnetic nanoparticles using arrays of MgO-based magnetic tunnel junction sensors*, Appl. Phys. Lett. *93*, 033903

Shi J, Mao X, Ahmed D, Colletti A, Huang TJ, **2008**: *Focusing microparticles in a microfluidic channel with standing surface acoustic waves (SSAW)*, Lab Chip *8*, 221-223

Sinha A, Ganguly R, Puri IK, **2009**: *Magnetic separation from superparamagnetic particle suspensions*, J. Magn. Magn. Mat. *321*, 2251-2256

Sivagnanam V, Song B, Vandevyver C, Gijs MAM, **2009**: *On-chip Immunoassay Using Electrostatic Assembly of Streptavidin-Coated Bead Micropatterns*, Anal. Chem. *81* (15), 6509-6515

Slonczewski JC, **1996**: *Current driven exertation of magnetic multilayers*, J. Magn. Magn. Mat. *159*, L1

SPRKKR, **2006**: *The Munich SPR-KKR package, version 2.2, by H. Ebert et al.*

Stoner EC, Wohlfarth EP, **1948**: *A Mechanism of Magnetic Hysteresis in Heterogeneous Alloys*, Phil. Trans. R. Soc. London A *240*, 599-642

Sudfeld D, Ennen I, Hütten A, Golla-Schindler U, Jaksch H, Reiss G, Meißner D, Wojczykowski K, Jutzi P, Saikaly W, Thomas G, **2005**: *Microstructural investigation of ternary alloyed magnetic nanoparticles*, J. Magn. Magn. Mat. *293*, 151-161

Sun S, **2006**: *Recent Advances in Chemical Synthesis, Self-Assembly, and Applications of FePt Nanoparticles*, Adv. Mat. *18*, 393-403

Sun Y, Li W. Dhagat P, Sullivan CR, **2005**: *Perpendicular anisotropy in granular Co-Zr-O films*, J. Appl. Phys. *97*, 10N301

Temam R, **1984**: *Navier Stokes Equations*, 3rd ed., North-Holland, Amsterdam

Thomasset F, **1981**: *Implementation of Finite Element Methods for Navier-Stokes Equations*, Springer, New York

Triebel H, **1980**: *Höhere Analysis*, 2nd ed., Deutsch, Thun, Frankfurt/Main

UMFPACK, **2009**, http://www.cise.ufl.edu/research/sparse/umfpack

Vadala ML, Zalich MA, Fulks DB, St. Pierre TG, Dailey JP, Riffle JS, **2005**: *Cobalt-silica magnetic nanoparticles with functional surfaces*, J. Magn. Magn. Mat. *293*, 162-170

Vieira G, Henigham T, Chen A, Hauser AJ, Yang FY, Chalmers JJ, Sooryakumar R, **2009**: *Magnetic Wire Traps and Programmable Manipulation of Biological Cells*, Phys. Rev. Lett. *103*, 128101

Wang S, Li G, **2008**: *Advances in Giant Magnetoresistance Biosensors With Magnetic Nanoparticle Tags: Review and Outlook*, IEEE Trans. Magn. *44*, 1687-1702

Wang Z, Lew WS, Bland JAC, **2006**: *Manipulation of superparamagnetic beads using on-chip current lines placed on a ferrite magnet*, J. Appl. Phys. *99*, 08P104

Weddemann A, **2006**: *Simulation des Bewegungszustandes magnetischer Partikel in Mikrokanälen*, Diploma thesis, Bielefeld University, Bielefeld

Weddemann A, **2008**: *Finite Elemente-Verfahren für ALE-Methoden*, Diploma thesis, Bielefeld University, Bielefeld

Whitesides GM, **2006**: *The origins and the future of microfluidics*, Nature *442*, 368-373

Wiklund M, Günther C, Lemor R, Jäger M, Fuhr G, Hertz HM, **2006**: *Ultrasonic standing wave manipulation technology integrated into a dielectrophoretic chip*, Lab Chip *6*, 1537-1544

Winter P, Athey P, Kiefer G, Gulyas G, Frank K, Fuhrhop, Robertson, Wickline S, Lanza G, **2005**: *Improved paramagnetic chelate for molecular imaging with MRI*, J. Magn. Magn. Mat. *293*, 540-545

Wittbracht F, **2007**: *Mikrofluidisches Sortiersystem für magnetische Beads*, Bachelor thesis, Bielefeld University, Bielefeld

Wittbracht F, **2009**: *Manipulation magnetischer Beads in Mikorfluidiksystemen: Separation und Positionierung*, Master thesis, Bielefeld University, Bielefeld

Woo K, Hong J, Ahn JP, **2005**: *Synthesis and surface modification of hydrophobic magnetite to processible magnetite@silica.propylamine*, J. Magn. Magn. Mat. *293*, 1797-1811

Yang CY, Lei U, **2006**: *Dielectric force and torque on a sphere in an arbitrary time varying electric field*, Appl. Phys. Lett. *89*, 163902

Die VDM Verlagsservicegesellschaft sucht für wissenschaftliche Verlage abgeschlossene und herausragende

Dissertationen, Habilitationen, Diplomarbeiten, Master Theses, Magisterarbeiten usw.

für die kostenlose Publikation als Fachbuch.

Sie verfügen über eine Arbeit, die hohen inhaltlichen und formalen Ansprüchen genügt, und haben Interesse an einer honorarvergüteten Publikation?

Dann senden Sie bitte erste Informationen über sich und Ihre Arbeit per Email an *info@vdm-vsg.de*.

Sie erhalten kurzfristig unser Feedback!

VDM Verlagsservicegesellschaft mbH
Dudweiler Landstr. 99   Telefon +49 681 3720 174
D - 66123 Saarbrücken   Fax   +49 681 3720 1749
www.vdm-vsg.de

Die VDM Verlagsservicegesellschaft mbH vertritt

Printed by Books on Demand GmbH, Norderstedt / Germany